Enhancing Life Cycle Reliability with Robust Engineering and Predictive Health Management

Wiley Series in Quality & Reliability Engineering

Dr. Andre V. Kleyner
Series Editor

The Wiley Series in Quality & Reliability Engineering aims to provide a solid educational foundation for both practitioners and researchers in the Q&R field and to expand the reader's knowledge base to include the latest developments in this field. The series will provide a lasting and positive contribution to the teaching and practice of engineering. The series coverage will contain, but is not exclusive to,

- Statistical methods
- Physics of failure
- Reliability modeling
- Functional safety
- Six sigma methods
- Lead-free electronics
- Warranty analysis/management
- Risk and safety analysis

Wiley Series in Quality & Reliability Engineering

Enhancing Life Cycle Reliability with Robust Engineering and Predictive Health Management
Matthew Hu and Yan-Fu Li
April 2026

Reliability Prediction for Microelectronics
Joseph B. Bernstein, Alain A. Bensoussan, Emmanuel Bender
March 2024

Software Reliability Techniques for Real-World Applications
Roger K. Youree
December 2022

System Reliability Assessment and Optimization: Methods and Applications
Yan-Fu Li, Enrico Zio
April 2022

Design for Excellence in Electronics Manufacturing
Cheryl Tulkoff, Greg Caswell
April 2021

Design for Maintainability
Louis J. Gullo (Editor), Jack Dixon (Editor)
March 2021

Reliability Culture: How Leaders can Create Organizations that Create Reliable Products
Adam P. Bahret
February 2021

Lead-free Soldering Process Development and Reliability
Jasbir Bath (Editor)
August 2020

Automotive System Safety: Critical Considerations for Engineering and Effective Management
Joseph D. Miller
February 2020

Prognostics and Health Management: A Practical Approach to Improving System Reliability Using Condition-Based Data
Douglas Goodman, James P. Hofmeister, Ferenc Szidarovszky
April 2019

Improving Product Reliability and Software Quality: Strategies, Tools, Process and Implementation, 2nd Edition
Mark A. Levin, Ted T. Kalal, Jonathan Rodin
April 2019

Practical Applications of Bayesian Reliability
Yan Liu, Athula I. Abeyratne
April 2019

Dynamic System Reliability: Modeling and Analysis of Dynamic and Dependent Behaviors
Liudong Xing, Gregory Levitin, Chaonan Wang
March 2019

Reliability Engineering and Services
Tongdan Jin
March 2019

Design for Safety
Louis J. Gullo, Jack Dixon
February 2018

Thermodynamic Degradation Science: Physics of Failure, Accelerated Testing, Fatigue and Reliability
Alec Feinberg
October 2016

Next Generation HALT and HASS: Robust Design of Electronics and Systems
Kirk A. Gray, John J. Paschkewitz
May 2016

Reliability and Risk Models: Setting Reliability Requirements, 2nd Edition
Michael Todinov
November 2015

Applied Reliability Engineering and Risk Analysis: Probabilistic Models and Statistical Inference
Ilia B. Frenkel (Editor), Alex Karagrigoriou (Editor), Anatoly Lisnianski (Editor), Andre V. Kleyner (Editor)
October 2013

Design for Reliability
Dev G. Raheja (Editor), Louis J. Gullo (Editor)
July 2012

Effective FMEAs: Achieving Safe, Reliable, and Economical Products and Processes Using Failure Modes and Effects Analysis
Carl Carlson
April 2012

Failure Analysis: A Practical Guide for Manufacturers of Electronic Components and Systems
Marius Bazu, Titu Bajenescu
April 2011

Reliability Technology: Principles and Practice of Failure Prevention in Electronic Systems
Norman Pascoe
April 2011

Improving Product Reliability: Strategies and Implementation
Mark A. Levin, Ted T. Kalal
March 2003

Test Engineering: A Concise Guide to Cost-Effective Design, Development and Manufacture
Patrick O'Connor
April 2001

Integrated Circuit Failure Analysis: A Guide to Preparation Techniques
Friedrich Beck
January 1998

Measurement and Calibration Requirements for Quality Assurance to ISO 9000
Alan S. Morris
October 1997

Electronic Component Reliability: Fundamentals, Modelling, Evaluation, and Assurance
Finn Jensen
November 1995

Enhancing Life Cycle Reliability with Robust Engineering and Predictive Health Management

Matthew Hu
Director, Master's Engineering Management Program
Associate Professor (Effective September 2026), Master's Engineering Management Program
Department of Industrial and Systems Engineering
University of Houston, Houston, USA
Member, Administrative Committee (AdCom), IEEE Reliability Society

Yan-Fu Li
Director, Institute for Quality and Reliability
Professor, Department of Industrial Engineering
Tsinghua University, Beijing, China
Associate Editor, Reliability Engineering & System Safety
Associate Editor, IEEE Transactions on Reliability (2017–2024)

WILEY

This edition first published 2026

Registered Offices
John Wiley & Sons, Inc., 111 River Street, Hoboken, NJ 07030, USA
John Wiley & Sons Ltd., New Era House, 8 Oldlands Way, Bognor Regis, West Sussex, PO22 9NQ, UK

For details of our global editorial offices, customer services, and more information about Wiley products visit us at www.wiley.com.

The manufacturer's authorized representative according to the EU General Product Safety Regulation is Wiley-VCH GmbH, Boschstr. 12, 69 469 Weinheim, Germany, e-mail: Product_Safety@wiley.com.

Wiley also publishes its books in a variety of electronic formats and by print-on-demand. Some content that appears in standard print versions of this book may not be available in other formats.

Library of Congress Cataloging-in-Publication Data Applied for

Hardback ISBN: 9781394182381

Cover Design: Wiley
Cover Image: © narvo vexar/Getty Images

Set in 9.5/12.5pt STIXTwoText by Straive, Chennai, India

To our families, mentors, and colleagues, whose unwavering support and inspiration have fueled our passion for advancing reliability and quality engineering. This book is dedicated to the innovators and practitioners who strive to create robust, reliable systems that enhance safety, efficiency, and trust in an ever-evolving technological world.

Contents

Series Editor's Foreword

The Wiley Series in Quality and Reliability Engineering aims to provide a solid educational foundation for researchers and practitioners in the field of quality and reliability engineering and to expand the knowledge base by including the latest developments in these disciplines.

The importance of quality and reliability to any engineering system can hardly be disputed. Product failures in the field inevitably lead to losses in the form of repair costs, warranty claims, customer dissatisfaction, product recalls, loss of sales, and, in extreme cases, loss of life.

Engineering systems are growing more complex as artificial intelligence (AI), autonomous vehicles, robotics, and deeper hardware–software integration push quality and reliability to the forefront of product development. Also, the continued miniaturization of integrated circuits—where feature sizes now reach just a few atomic layers—introduces new challenges for both design and manufacturing. Shorter development cycles only intensify the pressure, making robust design and product health management more critical than ever.

Robust design is aimed at making products and processes perform consistently as intended, regardless of variations in manufacturing tolerances, environmental conditions, user behavior, or component aging. This book treats reliability and robustness as a strategic mindset and presents a step-by-step methodology for implementing robust design in practice, covering the key tools needed to deliver a reliable product within today's compressed development timelines.

This book offers an excellent mix of theory, practice, useful applications, and common-sense engineering, making it a natural addition to the Wiley Series in Quality and Reliability Engineering. It advances the study of Physics of Failure, Prognostics and Health Management (PHM), and Design for Six Sigma (DFSS), strengthening our collective ability to meet the technological and engineering challenges ahead.

Despite its obvious importance, quality and reliability education remains critically underrepresented in engineering curricula. Few universities offer dedicated degree programs or even a meaningful range of courses in these methods, and topics such as Physics of Failure and Design for Reliability are largely absent across both undergraduate and graduate programs. As a result, most practitioners develop their expertise through colleagues, seminars, and technical publications rather than formal instruction—a gap that only deepens the profession's reliance on high-quality technical literature.

We are confident that this book, and the Series as a whole, will uphold Wiley's tradition of excellence in technical publishing and make a lasting contribution to the teaching and practice of reliability and quality engineering.

Dr. Andre V. Kleyner

Preface

The Critical Role of Quality and Reliability in Modern Engineering Systems

In today's fast-evolving technological landscape, the importance of quality and reliability in engineered systems cannot be overstated. Modern products—integrating advanced hardware, software, interconnected systems, rich electronics, and intricate subsystem interactions—are increasingly complex. Reliability requirements remain stringent or even intensify, as failures can lead to catastrophic consequences. For instance, the infamous sudden acceleration issues in Toyota vehicles severely damaged the company's profits and reputation, underscoring how a single failure can erode trust and cause profound implications.

Product failures in the field result in:

- **Financial losses:** Repair costs, warranty claims, product recalls, and heightened liability expenses.
- **Customer dissatisfaction:** Loss of trust, reduced market share, and declining revenue.
- **Operational disruptions:** Delays in next-generation development as teams resolve current issues, alongside persistent service demands from defective designs.
- **Reputational damage:** Negative publicity and diminished brand equity.
- **Safety risks:** Endangerment of human lives, especially in critical applications like autonomous vehicles, medical devices, aerospace systems, or IoT ecosystems.
- **Internal impacts:** Management distraction from future growth and declining employee morale and retention due to preventable failures.

To address these challenges, reliability engineering has shifted from a reactive discipline to a proactive, strategic function that drives innovation and competitive advantage. Viewing reliability as "quality over time"—the probability of survival without failure over a specified period, while maintaining consistent performance amid evolving stresses—positions it as a quality anchor. By embedding quality and reliability into the product development process, organizations can deliver robust systems that perform consistently under real-world conditions, sustaining customer satisfaction throughout the lifecycle.

At its core, product development is a risk management exercise. Failing to meet expectations for features, functionality, reliability, service life, or cost of ownership amplifies these risks. A disciplined, repeatable process is essential to capture institutional knowledge, reduce reliance on individual expertise, and ensure consistency across teams. This proactive mindset fosters innovation, mitigates risks, and aligns design decisions with long-term goals.

Reliability and Robustness: A Strategic Mindset

Robustness over time is not merely an engineering objective—it is a strategic mindset that shapes every stage of product development. Robustness over time is not just a task, but a way of thinking—one that drives action plans and execution from the earliest stages of product development, particularly in technology development. This philosophy transforms how teams approach design, shifting from isolated checklists or afterthoughts to an integrated, forward-looking perspective that permeates all decisions. By embedding robustness into the mindset from the start, teams can proactively identify potential risks, optimize design choices, and ensure the final product performs reliably under a wide range of conditions.

This mindset emphasizes:

- **Proactive risk identification:** Anticipating failure modes, stressors, and variability sources early in the concept phase, rather than waiting for issues to emerge during testing.
- **Cross-functional collaboration:** Engaging diverse teams—including engineering, manufacturing, quality, supply chain, and even end-user representatives—to explore design alternatives, share insights, and validate assumptions collaboratively.
- **Preventive focus:** Moving away from reactive failure mitigation in late-stage testing or post-launch fixes to systematic prevention during initial concept and design phases, thereby avoiding costly rework.
- **Real-world simulation and adaptation:** Accounting for variations in materials, processes, usage patterns, environmental stressors, and emerging technologies to ensure sustained performance, with built-in flexibility for future adaptations.
- **Action-oriented execution:** Translating the mindset into concrete action plans, such as milestone reviews focused on robustness metrics, iterative simulations, and cross-team workshops that prioritize long-term durability over short-term gains.
- **Cultural integration:** Fostering an organizational culture where robustness is a core value, encouraging continuous learning, knowledge sharing, and accountability at all levels to sustain this way of thinking across projects.

In technology-driven fields like AI, IoT, and autonomous systems, where innovation cycles are rapid and uncertainties abound, this mindset is particularly vital. It enables teams to navigate emerging challenges, such as evolving regulatory requirements or unforeseen integration issues, by building in margins for error and adaptability. Ultimately, by adopting robustness as a foundational way of thinking, organizations not only minimize risks but also unlock opportunities for differentiation, such as developing products that evolve with user needs or withstand technological disruptions. This approach ensures that robustness is a guiding principle from concept to launch, maximizing long-term performance, customer satisfaction, and market resilience.

The Modern Challenge: Balancing Speed, Reliability, and Robustness in Accelerated Development Cycles

The rapid advancement of technologies like autonomous driving, artificial intelligence (AI), Internet of Things (IoT), and advanced robotics has revolutionized product development. Reliability professionals must deliver innovative, high-performance products faster while exceeding customer expectations for durability, safety, and functionality. This pressure is intensified by the need to ensure robustness amid complex, unpredictable real-world conditions, often without physical prototypes.

Robustness—the ability to maintain stable performance despite variations in design, manufacturing, usage, and environmental factors—is foundational to long-term reliability. Key factors impacting robustness include:

- **Design parameter variations:** Differences in component tolerances, material properties, or specifications.
- **Material properties:** Inconsistencies in raw materials or degradation from wear, corrosion, or fatigue.
- **Use conditions:** Diverse user behaviors, operational patterns, and misuse scenarios.
- **Manufacturing processes:** Variations in production techniques, assembly methods, or quality control.
- **System interconnections:** Complex interactions between hardware, software, and networked components.
- **Environmental stressors:** Exposure to temperature extremes, humidity, vibration, electromagnetic interference, or other external factors.

These elements introduce variability that can compromise performance if unaddressed. Comprehensive robustness assessment and development provide a proactive framework to identify and mitigate risks early, reducing costly iterations and preventing failures in testing or post-launch.

A Structured Approach to Robustness Development: Integrating Six Sigma and IDOV

Achieving robust designs requires a structured, proactive framework. The robust design for reliability approach, often combined with Design for Six Sigma (DFSS), embeds robustness from the outset. Six Sigma provides a data driven methodology, rooted in the scientific method: observe problems, hypothesize, analyze data, and implement solutions.

Central to this are **critical-to-quality (CTQ)** metrics, which translate customer needs into measurable parameters, and **critical parameter management (CPM)**, which flows requirements down to subsystems while rolling up performance data for verification.

Six Sigma methodologies include:

- **DMAIC (Define–Measure–Analyze–Improve–Control):** Corrective for existing processes, eliminating defects.
- **DFSS:** Preventive, designing quality and reliability in from the start.

The **IDOV (Identify–Design–Optimize–Validate)** framework, aligned with DFSS, drives robust, error-free outcomes by integrating Six Sigma's rigor, DFSS's prevention, and reliability engineering's system perspective.

IDOV Phases:

1) **Identify:**
   - Define customer requirements and CTQs.
   - Identify variability sources (e.g., materials, use conditions, tolerances).
   - Establish performance and reliability goals.
2) **Design:**
   - Develop designs accounting for variations and stressors.
   - Use simulations to explore alternatives and assess robustness.
   - Incorporate redundancy, fault tolerance, or adaptive controls.
3) **Optimize:**
   - Based on robustness thinking, apply robust optimization methodologies like the Taguchi method or equivalent in the presence of noisy environments to minimize variability and maximize performance.
   - Identify critical design parameters and select them as potential health indicators for the assessment of PHM application through the whole lifecycle for time reliability.
   - Apply techniques like response surface methodology or genetic algorithms to further refine designs.
   - Balance trade-offs in cost, weight, reliability, and other constraints.
   - Employ Monte Carlo simulations to ensure designs remain robust across a range of scenarios.
4) **Validate:**
   - Conduct virtual testing with digital twins or high-fidelity simulations.
   - Perform targeted physical testing to confirm results.
   - Monitor health indicators for long-term stability.

This preventive focus shifts from reactive mitigation to systematic prevention, emphasizing cross-functional collaboration and real-world simulation.

Integrating Prognostic Health Management (PHM) for Enhanced Reliability

Prognostic health management (PHM) complements robustness by offering real-time system health insights. Monitoring indicators like performance degradation, failure precursors, and maintenance thresholds allow:

- Anticipating failures.
- Optimizing schedules to extend asset life.
- Feeding data back into designs for iterative improvements.

Linking PHM with Lean Six Sigma—through shared emphasis on CTQs, CPM, and critical factor monitoring—creates a closed-loop system. This integration ensures robust designs, proactive maintenance, and continuous improvement, optimizing performance while reducing lifecycle costs.

Overcoming Challenges in Accelerated Development

Accelerated cycles pose hurdles:

- **Time constraints:** Limited physical testing, heightening simulation reliance.
- **Complexity:** Intricate hardware–software network interactions.
- **Data availability:** Early-stage gaps requiring probabilistic models.
- **Coordination:** Need for cross-functional alignment.

Solutions include:

- Investing in advanced simulation tools for complex systems.
- Leveraging AI/machine learning for analysis and optimizations.
- Forming integrated product teams (IPTs) for goal alignment.
- Building data ecosystems to inform and validate designs.

The Strategic Importance of Robustness and Closing the Education Gap

Robustness is a strategic imperative, enabling:

- Reduced lifecycle costs via minimized claims and risks.
- Faster time-to-market by avoiding redesigns.
- Enhanced customer satisfaction and loyalty.
- Competitive differentiation through superior reliability.

In safety-critical industries, it distinguishes leaders from those facing recalls.

Yet, quality and reliability are undertaught in academia, with topics like warranty analysis or real-world modeling often overlooked. Professionals learn on-the-job or through resources. This book bridges the gap, offering practical guidance for graduate students, early-career engineers, and seasoned professionals.

About This Book

Authored by experts, this volume covers:

- System reliability and robustness modeling.
- Robustness assessment and optimization.
- Health indicators and prognostics.
- Integrating reliability, robustness, Six Sigma, and PHM.

Change Begins with You

Our aim in writing this book is to give you insights that are both valuable and practical—some may be entirely new; others may deepen what you already know. We've enjoyed the process of sharing the paradigms and principles that have consistently worked in our own experience. These strategies, tools, and practices are not passing fads or short-lived themes; they have stood the test of time and align well with today's push for leaner, more flexible, and more agile product development.

Of course, implementation evolves with experience. New methods take time to blend into an organization's broader set of practices, as teams learn how to adapt them for both immediate impact and lasting benefit. Keeping these approaches to ourselves would serve no one—our intent is for you to read, reflect, adapt, and apply them so you can build sustainable competitive advantage.

Whether you're an individual contributor or a manager, you have the power to influence outcomes:

- **As an individual contributor**, you can raise the quality of your work without waiting for a top-down mandate. Small, consistent improvements can ripple through the team, boosting both product quality and organizational efficiency.
- **As a manager**, you can set higher expectations for the quality of information used in decision-making. Better information means more than collecting data—it's about identifying what truly matters, defining relevant performance metrics, and tracking them closely. This clarity makes it possible to assess risks, solve problems effectively, and steer your team toward meaningful results.

The marketplace rewards those who execute with excellence. Strive to outpace your competition, serve your customers exceptionally well, and treat every improvement as a step toward long-term success.

Your journey starts now.

Conclusion

In an era of rapid advancement, robustness development and PHM are vital for reliable products. By leveraging simulations, structured methodologies like IDOV, and integrated approaches, organizations mitigate risks, innovate faster, and exceed expectations. Robustness remains the cornerstone of success, ensuring reliability, safety, and advantage in evolving industries. This book equips readers to design resilient systems meeting today's demands.

Matthew Hu
Yan-Fu Li

Acknowledgments

We owe our deepest gratitude to our mentors in robust design: Dr. Genichi Taguchi and his son, Shin Taguchi; Dr. Ranjit K. Roy; Prof. Kai Yang (Wayne State University); and Prof. K.C. Kapur (University of Washington). For their pioneering work in prognostic health management (PHM) and reliability engineering, we thank Prof. Way Kuo and Prof. Min Xie (City University of Hong Kong), Prof. Ming Zuo (University of Alberta), and Prof. Fugee Tsung (Hong Kong University of Science and Technology). Each has generously shared insights that have shaped our understanding of quality and reliability engineering.

We also appreciate the support of the University of Houston and Tsinghua University for enabling us to learn, teach, and guide others in applying these methods. Special thanks go to Mr. Ziying Li, an exceptionally talented and dedicated PhD student under Prof. Yan-Fu Li, whose contributions enriched several chapters of this book.

Our experience teaching undergraduate and graduate courses in the United States and China—centered on robust design and PHM in operations management, quality, and reliability—formed the foundation for much of this work. We are grateful to Gino Lim, current chairman of the Department of Industrial and Systems Engineering at the University of Houston, and Anthony Amber, former Dean of the College of Technology at the University of Houston, for their encouragement and support.

We thank Dr. Jimmy Hu Jianping, founder and chairman of Haylion Technologies, for the opportunity to apply our approach to autonomous technologies and vehicles in their early design phase. We also extend our appreciation to Meituan for partnering with us to implement PHM in the operation and maintenance of their advanced autonomous delivery vehicle fleet.

We are indebted to Jim Flanary, former President and General Manager of ITT Night Vision; Larry Smith (Ford Motor Company), former Quality Coach for the Heritage Team, who played a key role in revitalizing the Ford Rouge manufacturing complex; and Ann O'Neil, former Vice President of Quality and Lean Manufacturing at United Technologies Automotive, for their invaluable guidance and support.

While we cannot name everyone who has enriched this work, we extend our heartfelt thanks to all who have shared their experiences and expertise.

Finally, we thank our editor, Kavipriya Ramachandran; our technical reviewers, Sandra Grayson and Becky Cowan; and the entire Wiley team for their steadfast support.

1

Enchaining Lifecycle Reliability with Robust Engineering and Prognostic Health Management

CHAPTER MENU

1.1 Introduction

In a competitive market, quality, price, and cost of ownership are critical factors. The product's performance in its operational environment significantly impacts both the customer's quality experience and the cost of ownership. Initially, many products may attract buyers with their appealing features, but long-term success depends on customer-perceived reliability. If reliability falls short, negative sentiment and a damaged reputation can quickly arise and be difficult to rectify. From a producer's perspective, substantial amounts are often spent on warranties, with warranty costs sometimes equaling the product development expenses. Reliability is one of the most important characteristics of an engineering system. Reliability performance is largely established during the product development phase. Therefore, systematic efforts to prevent failures must be implemented throughout the entire development process, with specific strategies tailored to each phase. Poor product reliability typically stems from variation and a lack of knowledge. In this book, we present an approach for managing variation at various stages of product development. And we also introduce a new approach to addressing both lack of knowledge and variation within a unified framework.

Reliability can be measured as robustness over time, serving as a leading key performance indicator (KPI). Product development has a significant impact on revenue streams and reliability. It is most cost effective and less time consuming to make designs insensitive to uncontrollable user environments in terms of variation and uncertainty during the upfront design phase. Variation and uncertainty have become integral to our modern worldview. In the industrial sector, concepts like Six Sigma DMAIC (define, measure, analyze, optimize, control) and Design for Six Sigma (DFSS) have gained prominence. The quality movement initiated by Walter A. Shewhart was founded on understanding variation. Deming suggested that "understanding variation" should be one of the fundamental components of what he termed "profound knowledge," essential for management to significantly enhance organizational capability, as outlined in his 14 points for management. Shewhart introduced the important distinction between "chance causes" and "assignable causes" of variation. "Chance causes" stem from numerous minor, indistinguishable sources of variation, while "assignable causes" lead to noticeable changes in the process. According to Shewhart, only chance causes can be modeled using probability models, enabling probabilistic predictability. Processes exhibiting only chance causes are described as "in statistical control" and "predictable within limits." In manufacturing, the objective is to achieve processes under statistical control. However, in companies lacking systematic improvement programs, few processes achieve statistical control. Nonetheless, the outcomes of these processes are often modeled using probabilistic models, assuming that the underlying statistical processes are under some form of control and, therefore, predictable.

We aim to highlight the importance of understanding variation in creating reliable products. Additionally, we will broaden our perspective to discuss other uncertainties that may affect the reliability assessment of a product. The first type of uncertainty arises from random variation, the second type stems from our lack of knowledge, such as lack of good-quality data, when modeling product characteristics or estimating parameters.

How can we effectively manage variation and uncertainty? Our conventional approach may involve efforts to control variation and mitigate risks stemming from uncertainties. However, the challenge with this "control" mindset is that it can lead to solutions that are neither effective nor efficient, often resulting in extended implementation times and increased costs. Achieving a reliable design for a dynamic system involves ensuring it is robust and can withstand the influence of uncontrollable user environments, known as noise factors. Robustness refers to a state where a product is insensitive to these noise factors. Noise factors frequently contribute significantly to design failures and unsatisfied end users. In situations where there exists a model or prototype of a system or component, it may be possible to reduce the impact of noise factors by carefully selecting the levels of one or more design parameters.

Robustness thinking is essential to improve quality and reliability proactively by factoring the activities of design for reliability (DFR). Robustness development in manufacturing can reduce the variability of those processes with valuable benefits to manufacturing yields, cycle time, and costs. Robustness development in DFR process provides benefits in reduction of early-on physical testing and traditional test–fix–test cycles. Robustness thinking in prognostics and health management (PHM), which is utilized when high risks are associated with the mission-critical requirements affecting functionality, reliability, maintainability, and operability of the product, encourages innovative and cost-effective approach to plan for PHM. Robustness achieved early in development enables shorter cycle times in the later design phases. Robust design bridges the gap between product development, performance measurements in manufacturing and field tracking, and PHM strategy by identifying, selecting, and verifying health indicators in an earlier design phase, thereby improving the effectiveness of PHM.

1.2 Purpose

The purpose of this book is to build a robustness-thinking-based approach for robust design for reliability (RDfR) and prognostic health management as the best practice from early product design to the whole product lifecycle management in a high level of engineering confidence at the lowest cost and in shorter time. Our title for this book links two important engineering approaches: the development of robustness in a product's design as a leading indicator and is focused on preventing problems, and the integrative data fusion (i.e., design, manufacturing, field operation, and maintenance) for prognostic health management. As we will discuss, robustness is foundational to reliability, and its development should be the early focus of team members responsible for meeting reliability goals and tracking the health of lifecycle performance. Our plan is to provide value-added strategies for robustness development in new products and health management with helpfully characterized three main types of robustness development and reliability growth case study: intrinsic, instrumental, and collective. Applied appropriately, it can help generate information on how multiple forms of engineering knowledge come together to inform decision making within reliability contexts.

1.3 Essentials of Robustness and Robust Design in Reliability Improvement

An important development in reliability engineering is robust parameter design pioneered by Genichi Taguchi [1]. For any design concept, there is a potentially large space of control factor settings that will nominally place the function at the desired target value. In robust parameter design, the engineer explores the design space seeking changes that will make the system more robust while keeping the performance on target. Taguchi's method employs orthogonal arrays to explore the design space. At the same time, outer arrays or compounded noises are used to explore the range of possible operating conditions. Signal-to-noise ratio (S/N ratio) might be used as measures of the robustness of the system and guide the engineer to preferable levels of the control factors shown in Eq. (1.1).

$$\text{S/N ratio} = \frac{\text{Energy (or power) that is transformed into intended output}}{\text{Energy (or power) that is transformed into unintended output}}$$

$$\text{S/N ratio} = 10 \log \frac{\bar{y}^2}{\sigma^2} \tag{1.1}$$

Compounded noise is a noise management strategy, when system uncertainty is more complicated, like in a highly dynamic system in which the behavior varies over time. Robust design methods have continued to be refined and are still an active area of systems engineering innovation. To improve reliability, improving robustness in system, parameter, and tolerance design is essential.

There are three fundamental ways to improve the reliability of a product, manufacturing, and supply chain during the design stage [2–4]:

- Reduce the sensitivity of the product's function to the variation in the product parameters.
- Reduce the rate of change of the product parameters.
- Include redundancy.

Summary of essentials of robustness and robust design in reliability improvement is:

- Design-in reliability upfront.
- Eliminate failures prior to testing.
- Increase deployed reliability.
- Promote rapid, cost-effective deployment of health and usage monitoring systems.
- Improve diagnostic and prognostic techniques and processes.
- Decrease operational and support costs.

1.4 Effective Reliability Efforts in an Integrated Product Development Environment

System reliability is a key requirement for a system to function successfully under the full range of conditions experienced in industries, including oil and gas, automotive and others. From a probabilistic viewpoint, reliability is defined as the probability a system will meet its intended function under stated conditions for a specified period of time; therefore, to predict reliability, you must know three things:

- Function.
- Stated conditions.
- The specified useful life or time period.

A typical textbook that addresses reliability will present a set of probabilistic concepts, such as a survival function, failure rates, and mean times between failures. These concepts are related to a model of the causes of failure, such as component reliabilities or material and environmental variability. To quantify, specified operating conditions are defined as an agreed-upon range of allowable conditions or an estimated probability density function for uncertain or variable parameters. This approach is well suited to calculating predicted failure rates when all of the data are available.

To improve reliability prediction capability when useful data are not available or not sufficient, an alternative approach can be to:

- Identify all potential function failure modes, make a risk assessment, and implement countermeasures.
- Make the product insensitive to user environments.
- Identify shortfalls in verification test plans and enhance verification tests to ensure detection of all failure modes.
- Execute efficient verification tests that demonstrate a product is mistake free and robust under real-world use conditions.

System reliability requires fulfilling two critical conditions: mistake avoidance and robustness.

- Mistake, in this case, is defined as the error due to design decision and manufacturing operations. Examples of mistakes in product development include missing components, installing a component backward, or interpreting a software command as being expressed in inches when it is actually in centimeters. Product reliability can be improved by reducing the incidence of such mistakes through a combination of knowledge-based engineering and problem-solving processes, such as Six Sigma's DMAIC.
- Robustness is the ability of a system to function (that is, insensitive to the user's environment to avoid failure) under the full range of conditions that may be experienced in the field.

System design faces two different challenges:

- Developing a system that functions under tightly controlled conditions, such as in a laboratory.
- Making that system function reliably throughout its lifecycle as it experiences a broad set of real-world environmental and operating conditions.

An example of this real-world challenge is effective system reliability engineering. The most cost-effective and least time-consuming way to make a reliable product—one that is insensitive to the user environment, or robust—is to start in the development or design phase by discovering and preventing failure modes soon after they are created and implementing countermeasures before production.

Although most reliability professionals and companies already practice reliability engineering within design, whether formally or informally, implementing a structured reliability program can significantly improve business. Through our approach to lifecycle reliability through robustness development and PHM, we can help you reduce returns and warranty costs, enhance quality assurance, and comply with latest standards.

- Reliability concerns both robustness and mistake prevention.
- Reliability = Robustness + Mistake Prevention.

Proactively integrating Design for Reliability (DfR) through transfer function-based robustness improvement within the DFSS framework is essential for modern reliability engineering. Robust design is the foundational principle of reliability, enabling engineers to manage variability, protect margin, and prevent failures by design, not by post hoc correction. While traditional DfR practices have improved reliability, they are now largely procedural, test-centric, and reactive-an approach that is increasingly inadequate for today's complex, software-intensive, and AI-enabled systems operating under compressed development cycles. To address these challenges, Robust Design for Reliability (RDR) represents the modern evolution of DfR, emphasizing first-principle robustness thinking, upfront design optimization, and proactive failure prevention. This shift establishes reliability as a required product engineering deliverable, owned and designed by product engineering teams. Reliability engineers enable this outcome through methods and validation; however, reliability is created by design.

1.5 Enhancing Reliability Integration into the Product Development Process

One of the more significant challenges for reliability engineering in product development is the complete execution of the planned reliability process in product development. The reliability tools and techniques outlined in the product plan are often skipped or delayed, minimizing the opportunity for outputs to impact the program and product fully.

This contrast of planning and execution does not occur equally with product plan goals such as time to market, developed technical features, and product cost point. Each of those is closely measured in development and has a strong presence when resource is re-negotiated during the program. There is specific consideration that must be made for reliability due to its more complicated nature to measure performance and improve toward the target.

In its traditional formulation, reliability is stated as the probability of failure under specified operating conditions. A typical textbook that addresses reliability will present a set of probabilistic concepts such as a survival function, failure rates, and mean times between failures. These concepts are then related to a model of the causes of failure such as component reliabilities or material and environmental variability. To make the model quantitative, specified operating conditions are stipulated as an agreed-upon range of allowable conditions or an estimated probability density function for uncertain or variable parameters. This approach is well suited to calculating predicted failure rates once all the data are available. This is the general approach of texts that emphasize reliability analysis (such as Ushakov [5]) as well as texts oriented toward DFR (such as Rao [6]). This is a very sound approach, but here we present an alternative formulation of reliability that has proven very effective in the improvement of reliability early in the development of a new system.

1.6 Physics of Failure (PoF)

A development in reliability engineering closely related to this book is the PoF approach developed at the Computer Aided Life Cycle Engineering (CALCE) Electronic Products and Systems Center at the University of Maryland. The first instance in archival literature of the term "physics of failure" is in Pecht et al. [7], which emphasizes use of a physics-based model for reliability prediction and DFR. This approach has been extended to product development by [8] and to accelerated life testing by Kimseng et al. [9]. This study builds upon the conception of PoF and seeks to extend this conception to the earliest, creative phases of system design.

The PoF approach begins with the first stage of design. A designer defines product requirements, based on customer needs and the supplier's capabilities. These requirements can include the product's functional, physical, testability, maintainability, safety, and serviceability characteristics. At the same time, the service environment is identified, first broadly as aerospace, automotive, business office, storage, or the like, and then more specifically as a series of defined temperature, humidity, vibration, shock, and other conditions. The conditions are either measured or specified by the customer. From this information, the designer, usually with the aid of a computer, can model the thermal, mechanical, electrical, and electrochemical stresses acting on the product.

PoF is a scientific discipline that determines the root causes of failure for electrical, electronic, or electromechanical (EEE) items (Figure 1.1). These items can include:

- Piece parts
- Subassemblies
- Modules
- Equipment

Next, stress analysis is combined with knowledge about the stress response of the chosen design materials to identify where failure might occur (failure sites), what form it might take (failure modes), and how it might take place (failure mechanisms). Prior analyses of the root causes of actual field failures are of particular use at this stage. Failure is generally caused by one of the four following types of stresses: mechanical, electrical, thermal, or chemical, and it generally results either from the application of a single overstress or by the accumulation of damage over time from lower-level stresses. Once the potential failure mechanisms have been identified, specific failure mechanism models are employed. The validity of these models can be tested by conducting accelerated aging tests. If no models are available, or if the models are found to be inaccurate, then new models are developed using a series of statistically designed experiments, which identify the most important design and environmental factors governing failure and the mathematical relationship linking those factors to the time to failure.

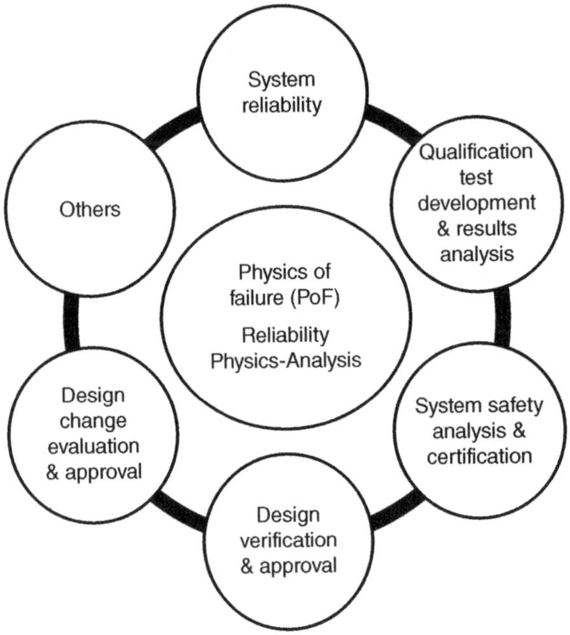

Figure 1.1 Reliability Physics Analysis.

Wherever possible, variabilities in the factors are specified using distribution functions. The effect of material properties, the geometry at failure sites, any damage to the part, and any manufacturing flaws and defects are addressed at this point as well. Once the models are established, a reliability assessment can be conducted on the product. This consists of calculating the time to failure for each potential failure mechanism, and then using the principle that a chain is only as strong as its weakest link, choosing the dominant failure site and mechanism as those resulting in the least time to failure.

The information from the assessment can be used to determine whether a product will survive for its intended application life, or it can be used to redesign a product for increased robustness against the dominant failure mechanisms and sites.

The PoF approach can be used to qualify design and manufacturing processes to ensure that the nominal design and manufacturing specifications meet or exceed reliability targets. Throughout the development cycle, it is employed to assess stress margins and establish process controls as a means of continuously improving the product's reliability from early prototype testing through final manufacture.

The PoF-based methodology is the preferred methodology for reliability assessment since it utilizes the knowledge of how failures actually occur in a product through degradation methods and can be used to evaluate a product's reliability based on its design and lifecycle. PoF ranks potential failure mechanisms and assists in the DFR, selection of stress test conditions, product qualification and screening, and performing PHM. Currently, PoF as a reliability assessment methodology is widely accepted by industry, professional societies, and the US government organizations, including the IEEE, EIA/JEDEC, and SEMATECH, which have been developing standards to incorporate PoF in the product development and evaluation process.

1.7 Failure-mode Avoidance

Thinking of reliability as failure-mode avoidance can have real advantages, especially in the early stages of system design or in long-term scenarios such as technology development. In early stages of system design, probability theory may be too quantitative for the task at hand. Probability density functions imply a level of precision in modeling the scenario that is often unwarranted, especially during early development. As a project advances through its development stages, the probabilistic view of reliability becomes increasingly useful. Analysis of reliability using probability theory is useful for component selection, system validation, and field-service operations. The value of the failure-mode avoidance conception of reliability is greatest for technology strategy, systems architecting, concept design, and for some robust parameter design activities, all done early during the development of the system.

An important development in reliability engineering is robust parameter design pioneered by Genichi Taguchi [10]. For any design concept, there is a potentially large space of control factor settings that will nominally place the function at the desired target value. In robust parameter design, the engineer explores the design space seeking changes that will make the system more robust while still keeping the performance on target. Taguchi's method employs orthogonal arrays to explore the design space. At the same time, outer arrays or compounded noises are used to explore the range of possible operating conditions. S/N ratios are used as measures of the robustness of the system and guide the engineer to preferable levels of the control factors.

Taguchi's philosophy of robust design is consistent with the approach to reliability engineering discussed here. Taguchi rejected the "goal post" mentality inherent in tolerance limits and specifications. His notion of a quality-loss function replaced consideration of defect rates and process yields with an emphasis on reducing variance followed by adjustment to target. Taguchi encouraged engineers to deliberately expose designs to harsh conditions in experiments. To do this requires a transformation in the culture of an engineering organization. The emphasis must shift from demonstrating adequate performance with high statistical confidence to aggressive improvement followed by adequate confirmation.

Another approach relevant to this study known as "operating window methods" was developed and practiced at Xerox Corporation in the 1970s. The operating window is the set of conditions under which the system operates without failure. In operating window methods, reliability is improved by making the operating window larger. Clausing [11] described the approach in detail in a recent issue of *Technometrics*, but the essence of the approach is simple enough to present here:

- Increase the value of the noise factors so that the failure rate is high.
- Change the value of the control factors to seek a broader operating window at a fixed failure rate.

Despite the use of failure rates as a measure of performance, the operating window method is, upon closer examination, consistent with Taguchi's quality philosophy. Because failure rates were greatly increased by applying aggressive noises, improvements could be made rapidly, even though they sacrificed the ability to accurately predict field reliability.

At the earliest stages of system design when our latitude to make changes is greatest, it is these types of conceptual changes that are most critical to find and implement. Although robust parameter design has been a valued development in systems engineering, large changes in system reliability observed over time cannot be explained by parameter design alone.

We promote and seek to help engineers implement early-stage robustness work, robustness thinking, and robust design methodologies. The next section covers some theoretical developments.

1.8 Design for Six Sigma

DFSS describes the application of Six Sigma tools to product development and process design. The goal is to "design in" Six Sigma performance capability. DFSS is an approach to designing (or redesigning) a product or service. It is equally useful in developing business processes or technical products. DFSS is a defined method—a culture and a way of viewing value creation.

The focus of DFSS begins with critical voice of customer (VOC) analysis and rational business planning. After gaining an understanding of the market and customer needs, design personnel work to understand and characterize critical design parameters and functionality. To achieve a cultural shift—focused on continuous improvement—you must go beyond DMAIC by leveraging a full suite of performance improvement tools. The time to develop new products is a critical success factor in almost any business today. DFSS helps reduce development time by deploying lessons learned throughout the development and manufacturing setup process.

DFSS provides many tangible benefits to organizations. For instance, the DFSS approach results in long-term cost reductions for a product. There are many ways these savings are realized. Instead of debugging products and processes that already exist, DFSS is a reexamination of the function and design parameters.

For a company that has achieved success through Six Sigma activities or equivalent—systematically applying statistical problem-solving methodologies to reduce variation and enhance quality—it is natural to extend these practices into the early stages of product and process development, known as DFSS. In essence, DFSS involves selecting product and process characteristics crucial to customers and manufacturing, identifying noise factors that influence these characteristics, and ensuring the product or process is resilient to variations in these factors. Both innovative approaches and systematic parameter design methodologies are employed.

DFSS starts from scratch with the goal of designing virtually error-free products or processes. This strategy effectively replaces the trial-and-error or built-test-fix processes, and results in product designs that consistently meet customer requirements. There are several different DFSS roadmap models:

- Invention, innovation, develop, optimize, and verify (IIDOV).
- Define, concept, design, optimize, and verify (DCDOV).
- Identify, define, develop, optimize, and verify (IDDOV).
- Define, measure, analyze, design, and verify (DMADV).
- Identify, design, optimize, and verify (IDOV).

Each has a different focus on generic technology development or product commercialization. The roadmap names are not important, but the contents and tasks at each phase defined to enhance product development process are.

A Typical DFSS Approach Includes Four IDOV Phases:

I—Initiate project. Initiate and prepare projects, establish business case, and project infrastructure.

D—Design development. Define requirements. Gather and understand customer requirements and develop concepts. Develop, evaluate, synthesize, and select design concepts.

O—Optimize design. Model, analyze, and develop design to optimize product and process capability.

V—Verify and validate. Test design outputs against customer requirements under defended operating conditions.

Robustness thinking and robust design are essential components of DFSS, aimed at enhancing product quality and process efficiency. These approaches focus on minimizing variability and ensuring that designs can withstand various conditions and stresses without failure. The ultimate goal is to eliminate inefficiencies and waste in both the design and production processes, thereby saving resources and aligning the end product or service more closely with customer needs, leading to higher satisfaction levels.

1.8.1 The Essence of Robustness Thinking

Robustness thinking is a **first-principle, design-centric approach to reliability engineering** that deliberately anticipates, manages, and absorbs variability and uncertainty before failures can occur. Its objective is not merely to respond to failures after they emerge, but to **prevent failures by intent through proactive design optimization**, rather than discovering them by chance through testing or field experience.

At its core, robustness thinking establishes reliability **upfront**, where design decisions have the greatest leverage and lowest cost of change.

Anticipation and Management of Variability

- Recognizing that **variability is inherent and unavoidable** in real-world systems, arising from manufacturing tolerances, environmental conditions, usage patterns, aging, and operational uncertainty.
- Treating variability and noise as **design inputs**, not exceptions.
- Actively identifying, characterizing, and managing sources of variation early in development to preserve performance margins and failure distance.

Proactive Design Optimization

- Optimizing designs for **stable performance across a wide operating envelope**, rather than maximizing performance under ideal or nominal conditions.
- Allocating and protecting **design margin** to ensure functionality under worst-case and off-nominal scenarios.
- Using transfer functions, sensitivity analysis, and parameter robustness to minimize the impact of noise factors on critical performance characteristics.

Focus on Performance Consistency

- Prioritizing **consistent and predictable system behavior** over peak or point-optimized performance.
- Ensuring that product performance remains within acceptable limits despite uncertainty, degradation, and external disturbances.
- Viewing consistency as a leading indicator of long-term reliability.

Proactive Failure Prevention

- Addressing potential failure mechanisms **during concept and design**, rather than reacting to failures during testing, validation, or field operation.
- Shifting reliability practice from **failure discovery and correction** to **failure avoidance by design**.
- Reducing dependence on extensive testing as a means of uncovering weaknesses late in the lifecycle.

Systems and Lifecycle Perspective

- Treating the product as part of an **integrated system**, encompassing design, manufacturing, supply chain, usage environment, maintenance, and end-of-life considerations.
- Ensuring robustness across the **entire lifecycle**, not just at launch.
- Enabling downstream activities—testing, PHM, digital twins, and AI—to **validate and monitor design intent**, rather than compensate for insufficient robustness.

Key Principle
Robustness thinking enables failures to be prevented deliberately by design, rather than discovered accidentally through testing or field exposure.

1.8.2 Robust Design as a Key Strategy

Robust design, also known as Taguchi methods, is a key strategy within robustness thinking. It aims to create designs that are insensitive to variations in manufacturing, environment, and usage. Key aspects include:

Parameter Design

- Optimizing design parameters to minimize the effect of noise factors on product performance.

Tolerance Design

- Determining the optimal tolerances for parameters to balance performance and cost.

Signal-to-Noise Ratio

- Using this metric to measure how well a design performs in the presence of noise factors.

Driving Proactive Reliability Efforts

Robustness thinking and robust design drive proactive reliability efforts by:

1) Early Identification of Potential Issues:
   - By considering variability and noise factors from the start, potential reliability issues are identified and addressed early in the design process.
2) Reduction of Sensitivity to Variation:
   - Robust designs are less sensitive to variations in manufacturing, usage, and environment, leading to more reliable products.
3) Emphasis on Prevention Rather than Detection:
   - The focus shifts from detecting and fixing failures to preventing them from occurring in the first place.
4) Integration of Reliability Considerations Throughout the Design Process:
   - Reliability becomes an integral part of design decisions rather than an afterthought.

1.8.3 Paradigm Shift and Change

The adoption of robustness thinking and robust design represents a significant paradigm shift in reliability engineering:

1) **From Reactive to Proactive:**
   - Instead of waiting for failures to occur and then fixing them, the focus is on preventing failures through robust design.
2) **From Ideal Conditions to Real-World Variability:**
   - The emphasis shifts from optimizing for ideal conditions to ensuring consistent performance across a range of conditions.
3) **From Component-level to System-level Thinking:**
   - Reliability is considered at the system level, including interactions between components and with the environment.
4) **From Quality Control to Quality by Design:**
   - Quality and reliability are built into the product from the start, rather than being inspected in at the end.
5) **From Cost Center to Value Creator:**
   - Reliability efforts are seen as creating value through improved customer satisfaction and reduced warranty costs, rather than as a necessary expense.

Robustness thinking and robust design are critical disciplines within DFSS that drive proactive reliability efforts and initiate a paradigm shift in product development. By focusing on minimizing variability and ensuring consistent performance under various conditions, these approaches enhance product quality and process efficiency. The ultimate goal is to eliminate inefficiencies and waste, align products more closely with customer needs, and achieve higher satisfaction levels. As organizations adopt these principles, they can expect to see significant improvements in product reliability, customer satisfaction, and overall market success.

1.8.4 DFSS Roadmap: Emphasizing Robustness

The DFSS roadmap is integral to facilitating robust design. The revised DFSS roadmap emphasizes robustness through its stages.

Case Studies and Applications

Real-world applications of robustness thinking and robust design can be seen across various industries. For example:

- Automotive Industry: Manufacturers use robust design principles to ensure that vehicles can withstand a wide range of operating conditions and provide reliable performance over their lifespan.
- Consumer Electronics: Companies design devices to be resilient against common user errors and environmental factors, ensuring consistent performance and customer satisfaction.
- Healthcare: Medical device manufacturers apply robust design techniques to create equipment that performs reliably in critical applications, enhancing patient safety and outcomes.

In every phase of the product development process, we provide nonbinding recommendations regarding the tools that could be used to achieve the necessary deliverables for progressing to the next phase. These recommendations are intended to support rather than restrict the creativity and imagination of the designer. Additionally, checklists are provided to ensure that all required outputs for each phase are captured. From a DFSS perspective, these checklists ensure that no critical aspects are overlooked or omitted.

1) **Identify**—Identify market needs. Define customer requirements and project goals. Identify critical to satisfaction (CTS) and related functional targets.

   Reliability is often a key CTS on the reliability aspects of a product. The purpose of this stage for the reliability effort is to clearly and quantitatively define the reliability requirements and goals for a product, as well as the end-user product environmental and use conditions.

   These can be at the system, assembly, component, or even the failure-mode level. Requirements can be determined in many ways or through a combination of those different ways. Requirements can be based on contracts, benchmarks, competitive analysis, customer expectations, cost, safety, and best practices. Some of the tools worth mentioning that help quantify the VOC include Kano models, affinity diagrams, and pair-wise comparisons. Of particular interest to DFR are the requirements that are critical to reliability (CTR).

   The system reliability requirement goal can be allocated to the assembly, component, or even the failure-mode level. After the requirements have been defined, they must be translated into design requirements and into manufacturing requirements.
2) **Design**—The purpose of the design phase in RDfR (IDOV) is to develop products that maintain high performance and dependability under a wide range of conditions and throughout their intended lifespan. Understand the system and select design concepts. Map CTS characteristics to lower-level y factors. Relate y factors to

critical to quality (CTQ) or CTR x design factors. Determining use and environmental conditions is an important early step of a DFR program. Know what it is to be designed for and what types of stresses the product should withstand.

The conditions can be determined based on customer surveys, environmental measurement, and sampling. The tendency for the potential failure-mode occurrence is aggravated by noise factors, which are those that engineers have little or no control and negatively influence designed system performance. Fundamental to designing for reliability and robustness using transfer function is the inclusion of noise factors during analysis that challenge the design and uncover potential failure modes.

After uncovering, these failure modes can be avoided by developing appropriate countermeasures—either in the design or manufacturing process. Including noise factors in upfront design analysis has encouraged engineers developing transfer functions to consider appropriate noise factors and realistic levels, as well as strategies to include them in simulations.

It is important to estimate the product's reliability, even with a rough first-cut estimate, early in the design phase. This can be done with estimates based on engineering judgment and expert opinion, PoF analysis, transfer-functions-based simulation models, prior warranty and test data from similar products and components (using life data analysis techniques), or standards-based reliability prediction.

3) **Optimize**—Design for robust and reliable performance. That minimizes product or process sensitivity to uncontrollable user environment to have better manufacturability and higher reliability.

   In this stage, a robust parameter design helps further factor reliability tasks into the design process by optimizing design function in the presence of noise factors to:

   - Identify important variables.
   - Estimate their effect on a certain product characteristic.
   - Optimize the settings of these variables to improve design robustness.

   Noise screen experiments may be necessary to identify high-impact noise factors to single out significant factor results in more realistic reliability tests and more efficient accelerated tests (because resources are not wasted on including insignificant stresses in the test) prior to the robust optimization efforts.

   Within the DFR concept, you are mostly interested in the effect of stresses on your test units. Robust design plays an important role in DFR because it assists in identifying the factors that are significant to the product's life, especially when the PoF are not well understood. The robustness of the given concept design can be used to assess the limitation of the given concept design from a reliability improvement perspective.

4) **Verify**—Assess the integrated system and subsystem effects on performance. Use reliability and manufacturing verification to assess design performance and the ability to meet customer requirements.

(Continued)

(Continued)

If the design has been "demonstrated," the product can be released for production. When reaching the manufacturing stage, the DFR efforts should focus primarily on reducing or eliminating problems introduced by the manufacturing process. Manufacturing introduces variations in material, processes, manufacturing sites, human operators, and contamination.

Because manufacturing piece-to-piece variation has been considered as part of noise factors and was optimized in the optimize phase, the product's performance should be insensitive to manufacturing variation if the noise factors were identified and incorporated in the optimize phase for the robustness study.

However, reliability may be reevaluated considering additional process variables. Design modifications might be necessary to improve robustness. For example, a design should require the minimal possible amount of nonvalue-added manual work and assembly. Whenever possible, it should use common parts and materials to facilitate manufacturing and assembling. It should also avoid tight design tolerances beyond the natural capability of the manufacturing processes.

1.9 Design for Reliability

DFR is a process. Specifically, DFR describes the entire set of tools that support product and process design (typically from early in the concept stage all the way through to product obsolescence) to ensure that customer expectations for reliability are fully met throughout the life of the product with low overall lifecycle costs. In other words, DFR is a systematic, streamlined, concurrent engineering program in which reliability engineering is woven into the total development cycle. The purpose of the DFR process is to provide requirements for DFR activities, which are intended to be an integral part of every product development effort to continuously improve product reliability and robustness. The reliability process integrates with a generic technology and product development process and can be tailored as specified in the technology and product development process. The product development process defines the scope and applicability.

The reliability plan documents the tailoring of the DFR activities. The reliability plan is created by the design team. It is the responsibility of the design team to implement the DFR by completing the activities outlined in this plan. The team must leverage a set of reliability engineering tools along with a proper understanding of when and how to use these tools throughout the design cycle. This process encompasses a variety of tools and practices and describes the overall order of deployment that an organization must follow to design reliability into its products. The reliability is part of the DFSS scorecard. DFR tasks can be well aligned with and embedded in a DFSS roadmap.

To make reliability a key product requirement and understand where reliability efforts stand in terms of the DFR process for designing and manufacturing for reliability, a DFR assessment scorecard can be helpful. The DFR assessment drives reliability goal setting, understanding the quality history, tool selection activities, testing strategies, and reliability demonstration through DFR gates review.

The DFR process can follow the DFSS roadmap—for example, the IDOV framework. With reliability in mind, product program teams can identify the boundary and scope of system requirements and design the product. Meaningful test progression strategies can be developed and emphasized through optimizing the design over the time domain and functional validation of the product.

DFR activities are part of various elements in technology and product development activities during the complete product lifecycle. Goals of the DFR process are as follows:

- Integrate VOC into product requirements to improve reliability and robustness of the product.
- Provide requirements for activities involved in the DFR/DFSS process. Optimize the design over the time domain and functional validation of the product using a test progression strategy.
- Identify methods for defining product reliability requirements and activities involved at each stage of product development.
- Provide the practitioner a means of prioritizing the reliability projects and studies that must be undertaken.
- Continuously improve product reliability and robustness over time.

DFR is an industry-wide practice and a philosophy of considering reliability in an early stage of product design and development, to achieve a highly reliable product with sustainable cost. PoF is recognized as a key approach to implementing DFR in a product design and development process. The author will present a case study to illustrate predicting and identifying product failure early in the design phase with the help of a quantitative PoF-model-based analysis tool.

1.10 Prognostics and Health Management

To fulfill the increasing demand on functionality and quality, modern systems are often built with overwhelming complexities. These systems often feature rich electronics and intricate interactions among subsystems/components. Extremely high requirements of system reliability are essential since a single failure can result in catastrophic consequences. Despite every effort made in the past, disasters keep occurring with profound implications. For example, the infamous sudden acceleration failures of Toyota automobiles which have significantly damaged the company's profit and reputation [8].

In view of the high impact and extreme costs usually associated with system failures, methods that can predict and prevent such catastrophes have long been investigated. Applications of developed methods are not rare in domains such as electronics-rich systems, aerospace industries, automotive industry, or even public health environment [2, 12]. In general, these methodologies can all be grouped under the framework of PHM. Particularly, prognostics is the process of predicting the future reliability of a product by assessing the extent of deviation or degradation of the product from its expected normal operating conditions; health management is the process of real-time measuring, recording, and monitoring the extent of deviation and degradation from normal operation condition [13, 14]. Different from traditional handbook-based reliability prediction methods (e.g., US Department of Defense Mil-Hdbk-217 and Telcordia SR-332 (formerly [15])), which assumed that constant hazard rate of each component can be tailored by independent "modifiers" to account for various quality, operating, and environmental conditions, PHM methodologies instead monitor the health state in real time and dynamically update the reliability function (hazard rates) based on in situ measurements and tailored evolution models obtained from historical data. Due to the success of existing PHM methodologies, there is growing interest in studying new PHM techniques and applying PHM to underdeveloped domains.

Nevertheless, the increasingly complex modern systems pose new challenges on PHM. One of the most prominent problems is called no fault found (NFF) problem (related terminologies include "cannot duplicate," "re-test OK," "trouble not identified," and "intermittent malfunctions"), particularly in electronic-rich systems. As the name suggests, it refers to the situation that no failure/fault can be detected/replicated during laboratory tests even when the failure has been reported in the field. NFF issues not only make the prognosis and diagnosis extremely difficult, but also can cause skyrocketing maintenance cost. NFF contributes a lot in operational cost in many different application areas. On top of the maintenance costs, potential safety hazards related to NFFs are even more striking. For example, both Toyota and National Highway Traffic Safety Administration (NHTSA) spent quite a long time to investigate the root causes of sudden acceleration failures in some car models, a problem that might be linked to 89 deaths in 71 crashes since 2000 according to NHTSA [16]. Unfortunately, no conclusive finding has been reached despite the efforts in trying to repeat the failures in a variety of laboratory conditions. These intermittent faults are also suspected to be the main reason for other catastrophes.

Intermittent faults or NFF problems pose significant barriers in applying traditional methods for reliability prediction, which are often empirical, and population based. From the examples, we can find that intermittent faults are often tightly related to the environmental conditions and operation histories of the individual system. They can hardly be repeated due to unknown random disturbances involved. Therefore, laboratory testing and assessment can only provide a reference on the "average" characteristics of the whole population and are insufficient to provide accurate modeling and prediction for everyone. To reduce the maintenance cost and eliminate safety hazards caused by NFF, robustness-thinking-based product design and manufacturing and the paradigm of PHM need to shift from empirical to data supported and from population based to individual based.

Along with the challenges, the fast development of information and sensing technology has enabled the collection of many in situ measurements during operations and provided the capability of real-time data management and processing for everyone. These advancements provide us great opportunity to develop sophisticated models with increasing accuracy of prognostics for individual items. For instance, many different types of data during the whole lifecycle of the products can be easily retrieved, especially in critical applications. These data could include product design, verification and validation tests, production process information, quality records, operation logs, and sensor measurements.

Businesses including companies of automotive, wind energy, oil and gas, engineering, manufacturing, and others are constantly challenged to reduce time to market, reduce product development cost, and reduce production costs by increasing manufacturing productivity while maintaining product quality and reliability. Hence, PHM technology has become of particular interest for industries. It allows maintaining the high standard of operational performance by detecting impending failures, while increasing equipment availability through a more effective maintenance scheduling supported by a prognostic capability. PHM permits the reliability of a system to be evaluated in its actual lifecycle conditions, to determine the advent of failure, and mitigate the system risks. Hence, hence health assessment, as a part of diagnostic, is the process of assessing the current health of a system. Data-driven methodologies in PHM are closely related with those in some other major research directions, such as statistical quality control, reliability engineering, and robust engineering. It is worthwhile to briefly discuss their relations with PHM.

Statistical quality control is an area that has been extensively studied for many decades. The main objective is to detect abnormalities or changes in a process. It is generally applied to many homogeneous units and focuses on identifying the abnormal ones which may be traced back to process faults. PHM, on the other hand, focuses more on how faults happen and how to predict future faults so that optimal maintenance policy can be made, rather than fault detection.

The research in PHM is closely related to those in reliability engineering, such as failure prediction and maintenance. Many methods in PHM stem from those originally developed in reliability engineering. However, they have different focuses of interests. Traditional reliability engineering focuses on the modeling and prediction of the entire product population, without much emphasis on variability of the individuals and their respective working conditions. Therefore, reliability engineering is most valuable for manufacturers' product design and warranty policy making where population characteristics are crucial, while PHM is most valuable for end users who care more about the specific units they have on hand.

Compared to PHM, which emphasizes online monitoring and dynamic updating as indicated in Chapter 5 (Section 5.2.3.1: Activities of Prognostics and Health Management (PHM) and Reliability Over the Product Lifecycle), robust engineering, from an engineering physics perspective, focuses on developing conceptual design robustness and further optimizing robustness in detailed design, such as parameter design and tolerance design, during the early design stage. PHM and robust design are implemented at different stages. Robust engineering is applied mostly during the system planning, concept design, and detail design phase, instead of its operating phase where PHM is applied. Robust design focuses on improving the fundamental function of the product or process, thus facilitating flexible designs and concurrent engineering. Robust design emphasizes on desensitizing the harmful side effects of noise (referred to as uncontrollable user environment). When variability occurs, Taguchi said this is because the physics active in the design and environment promotes change [17]. Noise environment can be categorized into five categories:

1) Piece-to-piece variation, such as rubber thickness.
2) Change over time, such as failure from material wear, or changes in force or dimension with time.
3) Customer use, such as open-hole wellbore size.
4) Environmental conditions, such as temperature variation.
5) System interfaces and interactions, such as elements outside dimension variations and open-hole size.

The result of noise may be degradation in quality (soft failure) or a malfunction failure (hard failure). A product is said to be robust when it is insensitive to the effects of sources of variability, even though the sources themselves have not been eliminated. The power of integrated robust engineering and PHM is an engineering disciplined approach that provides a framework and discipline to minimize or eliminate potential birth defect in concept design, strengthen product health conditions through robust optimization in detail design, including parameter design and tolerance design. Critical parameters (both products and processes) identified as health indicators during robust design and optimization are excellent health indicators for PHM in its operating phase applications such as manufacturing, maintenance, field, and others. These identified health indicators have much higher engineering confidence guiding conventional reliability engineering and PHM activities.

In case of data-driven methods, there is a need to identify and extract indicators (i.e., precursors to failure) through raw data gathered from the process. As a prior event analysis, prognosis is to predict whether a fault is impending and estimate how much time remains before a likely failure occurs using dynamics of these indicators. Although numerous PHM methodologies have been proposed with application on data from different areas, a few studies have recently investigated PHM regarding application in health indicators identification. However, identification of health indicators through robust engineering has better engineering confidence and is proactive and predictive.

The goal of health assessment is to provide information on the current condition of an operation system (i.e., its degradation state). Parameters corresponding to degradations can be detected through collected data, but the states of system are usually difficult to observe directly. The most common indicator of a proved failure state is the production of out-of-specification products. System's health assessment thus implies the integration of methodologies to extract from the original set of data relevant indicators of a system's critical characteristics. The availability of many different data types throughout a manufacturing process allows the development of several types of indicators, with the combination of multiple data-driven methods.

PHM is a technology to enhance effective reliability and availability of a product in its lifecycle conditions by detection of current and approaching failures by providing for mitigation of the system risks and is a maintenance policy aimed at predicting the occurrence of a failure in components and consequently minimizing unexpected downtimes of complex systems and have been studied by many researchers from many different engineering fields to increase system reliability, availability, safety, and to reduce the maintenance cost of engineering assets. Many works conducted in PHM research concentrate on designing robust and accurate models to assess the health state of components for applications to support decision-making. Prior knowledge to implement PHM in complex systems is crucial to building highly reliable systems. PHM implementation steps consist of:

- critical component analysis,
- appropriate sensor selection for condition monitoring (CM),
- prognostics feature evaluation under data analysis, and
- prognostics methodology and tool evaluation matrices.

PHM systems are some of the main protagonists of the Industry 4.0 revolution (Biggio and Kastanis [18]). Efficiently detecting whether an industrial component has deviated from its normal operating condition or predicting when a fault will occur is the main challenge these systems aim at addressing. Furthermore, they could potentially and drastically reduce the often-conspicuous costs associated with scheduled maintenance operations. The increasing availability of data and the stunning progress of machine learning (ML) and deep learning (DL) techniques over the last decade represent two strong motivating factors for the development of data-driven PHM systems.

PHM is an engineering field whose goal is to provide users with a thorough analysis of the health condition of a system and its components [12]. To this extent, PHM employs tools from data science, statistics, and physics in order to detect an eventual fault (anomaly detection) in the system, classify it according to its specific type (diagnostic), and forecast how long a system, e.g., machine, will be able to work in presence of this fault (prognostic) (Kadry [19]).

1.10.1 Health Indicators in Prognostic Health Management: Critical-to-Quality (CTQ) and Critical-to-Reliability (CTR)

In PHM, health indicators play a crucial role in predicting and mitigating potential failures and ensuring the reliability and quality of products throughout their lifecycle. Among these health indicators, CTQ and CTR parameters are paramount. Identifying and selecting these parameters are best achieved through RDfR, ensuring that products not only meet quality standards but also perform reliably under various conditions.

1.10.2 Critical-to-Quality (CTQ) Parameters

CTQ parameters are essential attributes that directly affect the quality of a product as perceived by customers. These parameters are derived from customer requirements and specifications and are critical for ensuring that the product meets or exceeds customer expectations. CTQ parameters can include factors such as performance, durability, aesthetics, and user-friendliness.

Significance of CTQ Parameters

- Customer Satisfaction: Ensuring that CTQ parameters are met is crucial for customer satisfaction and retention.
- Market Competitiveness: High-quality products that meet CTQ standards are more competitive in the market.
- Brand Reputation: Consistently meeting CTQ standards enhances brand reputation and trust.
- Regulatory Compliance: Adhering to CTQ parameters ensures compliance with industry standards and regulations.

1.10.3 Critical-to-Reliability (CTR) Parameters

CTR parameters are attributes that are essential for the reliable performance of a product over its expected lifespan. These parameters focus on the product's ability to function under various conditions without failure. CTR parameters include factors such as mechanical integrity, thermal stability, and resistance to wear and tear.

Significance of CTR Parameters

- Product Longevity: Ensuring that CTR parameters are met increases the lifespan of the product.
- Operational Safety: Reliable products reduce the risk of failures that could lead to safety hazards.

- Maintenance Costs: Products that meet CTR standards require less maintenance and have lower lifecycle costs.
- Customer Trust: Reliable products build customer trust and loyalty, leading to repeat business.

1.10.4 Identification and Selection of CTQ and CTR Parameters

The identification and selection of CTQ and CTR parameters are critical steps in ensuring product quality and reliability. These steps are best achieved through RDfR, a methodology that emphasizes designing products to perform reliably under various conditions.

1.10.4.1 Robust Design for Reliability (RDfR)

RDfR involves designing products that can withstand a wide range of operating conditions and variations. It focuses on minimizing the impact of variability on product performance, ensuring that critical parameters are consistently met. RDfR incorporates principles from robust design and reliability engineering to achieve these goals.

Steps in Identifying and Selecting CTQ and CTR Parameters

- VoC Analysis: Collect and analyze customer feedback to identify key quality and reliability requirements.
- Requirement Clarification: Define clear and measurable requirements for CTQ and CTR parameters based on customer needs and industry standards.
- Risk Assessment: Conduct risk assessments to identify potential failure modes and their impact on CTQ and CTR parameters.
- Statistical Analysis: Use statistical methods to analyze data and identify critical parameters that significantly affect quality and reliability.
- Design of Experiments (DOE): Conduct experiments to test the impact of various factors on CTQ and CTR parameters and identify optimal design solutions.
- Simulation and Modeling: Use simulation tools to model product performance and predict the impact of variability on CTQ and CTR parameters.
- Validation and Testing: Validate the identified CTQ and CTR parameters through rigorous testing and real-world trials.

1.10.5 Health Indicators in Prognostics and Health Management

In PHM, health indicators, including CTQ and CTR parameters, are used to assess the health of a product and predict potential failures. By continuously monitoring these indicators, PHM systems can provide early warnings of potential issues, allowing for proactive maintenance and risk mitigation.

Implementation of Health Indicators in PHM

- Sensor Integration: Integrate sensors to continuously monitor CTQ and CTR parameters in real time.
- Data Analytics: Use advanced data analytics to process sensor data and identify trends and anomalies.
- Predictive Modeling: Develop predictive models to forecast potential failures based on health indicators.

- Maintenance Scheduling: Use predictive insights to schedule maintenance activities and prevent unexpected failures.
- Feedback Loop: Create a feedback loop to update design and manufacturing processes based on insights from PHM systems.

The identification and selection of CTQ and CTR parameters are crucial for ensuring the quality and reliability of products. RDfR provides a structured approach to identify and manage these critical parameters, ensuring that products perform consistently under various conditions. Integrating these parameters as health indicators in PHM systems allows for proactive maintenance and risk mitigation, enhancing product lifecycle safety and reliability. By continuously monitoring and managing CTQ and CTR parameters, organizations can achieve higher customer satisfaction, operational efficiency, and competitive advantage in the market.

Digitalization is a fundamental driver of Industry 4.0 and Supply Chain 4.0, a novel paradigm which enhances production efficiency through information and communication technologies. These technologies also provide the foundation for Predictive PM for industrial components and systems, whereby CM data is employed to perform three tasks:

1) Detection of abnormal states, by identifying deviations from normal operating conditions in production processes, manufacturing equipment, and products.
2) Diagnostics, by classifying abnormal states.
3) Prognostics, by predicting the evolution of abnormal states up to failure.

In response to the emerging challenges as well as opportunities, taking advantage of robust design, as we discussed in this chapter on reliability earlier, to identify key product characteristics or system health characteristics or health indicator and determine appropriate PHM based on the findings from risk assessment, e.g., design failure mode and effect analysis (DFMEA), we provide an approach with best practices in this book to bridge the gaps, which may be leading to further improvement of lifecycle reliability through robustness development and PHM.

1.11 The Importance of Digital Quality in Lifecycle Reliability Through Robustness Development and Predictive Health Management

Digital quality is a transformative approach that leverages advanced digital technologies to enhance the reliability of products throughout their entire lifecycle. Integrating digital quality with robustness development and PHM ensures that products are not only designed for robustness but also maintained at optimal performance levels throughout their lifespan.

This comprehensive strategy significantly enhances lifecycle reliability, reduces downtime, and improves overall product performance and customer satisfaction.

1.12 Digital Quality in Lifecycle Reliability

Digital quality refers to the use of digital tools, technologies, and processes to ensure that products meet high standards of quality and reliability throughout their lifecycle.

This involves a combination of advanced analytics, simulation and modeling, real-time monitoring, and data-driven decision-making to enhance product robustness and reliability.

Advanced Data Analytics

- Implement big data analytics to identify patterns and predict potential reliability issues across the entire product lifecycle.
- Use ML algorithms to optimize not just design and manufacturing, but also operational performance and maintenance schedules.
- Develop real-time monitoring systems that track product performance from production through end of life.

Digital Twin Technology

- Create digital twins that evolve with the product, from design through disposal.
- Use digital twins for predictive maintenance, reliability optimization, and lifecycle management.
- Integrate digital twin data with product lifecycle management systems for a holistic view of product performance.

IoT and Connected Products

- Implement Internet of Things (IoT) technologies to gather real-time data on product performance throughout its operational life.
- Use connected product data to inform not just future design improvements, but also current product maintenance and upgrades.
- Develop PHM capabilities based on comprehensive IoT data.

Blockchain for Quality Assurance

- Implement blockchain technology for secure and transparent quality records across the entire product lifecycle.
- Use smart contracts to automate quality control processes from production through maintenance and disposal.
- Ensure the integrity and traceability of quality-related data throughout the product's life, including maintenance and repair records.

1.12.1 Robustness Development

Robustness development focuses on designing products that can perform reliably under a wide range of conditions and stresses. It involves identifying potential failure modes and implementing design strategies to mitigate these risks.

Advanced Simulation and Modeling

- Computer-aided Design (CAD) and Engineering (CAE): Utilize CAD and CAE tools to create detailed digital models of products. These models are used to simulate different operating conditions and identify potential failure points.

- Finite Element Analysis (FEA): Apply FEA to analyze the physical behavior of products under various loads and stresses. This helps in optimizing the design for robustness.
- Multi-physics Simulations: Implement multi-physics simulations to understand complex interactions within products as they age and undergo maintenance.

Failure Mode and Effects Analysis (FMEA)

- Systematic Risk Assessment: Conduct FMEA to identify potential failure modes, their causes, and their effects on the system. This allows for proactive design modifications to enhance robustness.
- Mitigation Strategies: Develop and implement mitigation strategies based on FMEA findings to address identified risks.
- Digital FMEA: Implement digital FMEA tools that can be updated in real time based on field data and performance insights.

Robust Design Techniques

- Robust Design for Reliability (RDfR): Incorporate RDfR principles to ensure that products are designed based on robust design principles with reliability as a key objective.
- Tolerance Design: Implement tolerance design to ensure that product components fit and function together reliably, even with variations in manufacturing processes.
- Design for Manufacturability and Assembly (DFMA): Integrate DFMA principles to ensure that products are not only robust but also easy to manufacture and assemble consistently.

Augmented and Virtual Reality in Design

- Use AR/VR technologies for virtual prototyping, design reviews, and maintenance planning.
- Implement AR-assisted assembly, maintenance, and repair processes.
- Develop VR training programs for quality control, maintenance, and end-of-life procedures.

1.13 Predictive Health Management (PHM)

PHM is a proactive approach that uses real-time data, advanced analytics, and ML to monitor and predict the health of products and systems. This allows for timely maintenance and repairs, reducing downtime, and extending the product's operational life.

Real-Time Monitoring and Data Collection

- IoT Sensors: Deploy IoT sensors to continuously monitor the condition of products and collect real-time data on various parameters such as temperature, vibration, and load.
- Data Logging: Implement data-logging systems to store and manage the collected data for further analysis.
- Edge Computing: Implement edge-computing solutions for real-time data processing and rapid response to potential health issues.

Advanced Analytics and Machine Learning

- Predictive Analytics: Use predictive analytics to analyze historical and real-time data to predict potential failures and maintenance needs. This helps in scheduling preventive maintenance before issues arise.
- Machine Learning Algorithms: Develop ML models to identify patterns and trends in the data that indicate potential reliability issues. These models improve over time with more data and usage.
- Digital Twin Integration: Use digital twins to compare actual product performance with expected behavior and predict future performance.

Condition-based Maintenance (CBM)

- Dynamic Maintenance Schedules: Implement CBM to adjust maintenance schedules based on the actual condition and performance of products rather than fixed intervals. This ensures maintenance is performed only when needed, reducing unnecessary downtime and costs.
- Proactive Repairs: Conduct proactive repairs and replacements based on predictive insights to prevent unexpected failures.
- Automated Maintenance Workflows: Develop automated systems that trigger maintenance workflows based on PHM insights, ensuring timely and efficient maintenance actions.

Cybersecurity in PHM

- Integrate cybersecurity considerations into the PHM process from design through disposal.
- Implement secure-by-design principles in connected products that remain robust throughout the product's life.
- Develop processes for ongoing security updates and vulnerability management throughout the product's operational life.

1.13.1 Integration of Digital Quality, Robustness Development, and PHM

Lifecycle Reliability Management

- Holistic Approach: Integrate digital quality, robustness development, and PHM into a holistic lifecycle reliability management strategy. This ensures that reliability is considered from the initial design phase through to end-of-life disposal.
- Continuous Improvement: Implement continuous improvement processes to regularly review and enhance reliability strategies based on feedback and new data.
- Adaptive Design Processes: Develop design processes that can quickly incorporate insights from PHM and field data to enhance product robustness in future iterations.

Collaborative Platforms

- Integrated Digital Platforms: Use integrated digital platforms to facilitate collaboration between design, engineering, manufacturing, and maintenance teams. This ensures that all stakeholders have access to the latest data and insights.

- Cloud-based Solutions: Leverage cloud-based solutions to store and share data across the organization, enhancing accessibility and collaboration.
- Cross-functional Data Analytics: Implement analytics tools that can process and analyze data from various sources (design, manufacturing, field use) to provide comprehensive reliability insights.

Customer Feedback and Insights

- User Data Analysis: Collect and analyze user data to gain insights into how products are used in real-world conditions. This helps in identifying areas for improvement and enhancing future designs.
- Customer Engagement: Engage with customers to gather feedback and address reliability issues promptly. This fosters customer satisfaction and loyalty.
- Predictive Customer Support: Use PHM insights to proactively engage with customers before issues arise, offering preventive maintenance or upgrades.

Supply Chain Integration

- Extend RDfR principles to suppliers and manufacturing partners to ensure robustness throughout the supply chain.
- Implement blockchain technology for supply chain transparency and traceability.
- Develop collaborative PHM strategies that include key suppliers and partners.

By implementing this comprehensive approach to digital quality, robustness development, and PHM, organizations can significantly enhance the lifecycle reliability of their products. This integrated strategy not only improves initial product quality and reliability but also provides valuable insights for continuous improvement, innovation, and effective health management throughout the product's lifecycle. The result is improved product performance, reduced downtime, significant cost savings, and higher customer satisfaction, ultimately leading to a stronger competitive position in the market.

1.14 Critical Parameter Development and Management (CPD&M): A Comprehensive Overview

Critical parameter development and management (CPD&M) is a systematic approach to product development and lifecycle management. It is a cornerstone of modern systems engineering, focusing on identifying, analyzing, and managing the most crucial parameters that govern the functionality, performance, and reliability of engineered systems. CPD&M is not just a development tool; it is a comprehensive strategy that spans the entire product lifecycle, from initial concept to post-launch management and even end-of-life considerations.

Fundamental Concepts of CPD&M

1) Definition and Scope: CPD&M is a structured process that identifies and manages parameters critical to a product's performance, reliability, and quality. These parameters, known as critical parameters, are those that pose significant risks during

development, manufacturing, packaging, maintenance, and throughout the product's lifecycle.

2) Cross-functional Collaboration: At its core, CPD&M is a collaborative effort involving diverse teams of engineers and technical specialists. This cross-functional approach ensures a holistic view of the product, considering all aspects from design to manufacturing and beyond.
3) Lifecycle Management: CPD&M extends beyond the development phase, encompassing the entire product lifecycle. This comprehensive approach ensures that critical parameters are managed effectively from conception to end-of-life, maximizing product performance and reliability.

1.14.1 The CPD&M Process

1.14.1.1 Initial Parameter Identification

The process begins with a comprehensive analysis of all design parameters. From this extensive set, the team identifies those parameters that pose the highest risk to product performance—these are designated as critical parameters.

1.14.1.2 The Seven Metrics

When a parameter is suspected to be critical, it undergoes rigorous evaluation based on the following seven metrics:

- **Measurability:** Assessing how accurately and consistently both input (X) and output (Y) variables can be measured.
- **Stability:** Evaluating the consistency of Y and controlling X variables over time.
- **Adjustability:** Examining how effectively Y can be adjusted by manipulating specific X variables across desired performance ranges.
- **Independence, Interactivity, and Statistical Significance:** Analyzing how X variables interact and their individual and combined effects on Y variables.
- **Hyper-sensitivity:** Measuring the degree of responsiveness between Y variables and their controlling X variables under nominal conditions.
- **Robustness:** Assessing how well Y variables maintain performance in the face of unwanted variations, particularly under stressful conditions.
- **Capability (Cp and Cpk):** Evaluating the process capability under both nominal and stressful conditions.

1.14.2 Continuous Improvement

Parameters that fail to meet all seven metrics satisfactorily are designated as critical and managed meticulously. The goal is to improve their performance, understand their context within the design, and potentially remove them from the critical list.

Applications of CPD&M

Proactive Development

- CPD&M can be applied proactively during technology and product/process development, ensuring that critical parameters are identified and managed from the earliest stages of design.

Reactive Management

- In post-launch scenarios, CPD&M can be used to optimize production processes and manage the product throughout its lifecycle.

Reverse Engineering

- CPD&M can be employed to back-derive an existing design's governing parameters, isolating critical parameters for cost reduction and lifecycle management decisions.

Integration with Other Methodologies

Prognostics and Health Management (PHM)

- Critical parameters identified through CPD&M serve as key health indicators in PHM systems, enabling predictive maintenance and risk management strategies.

Digital Quality Implementation

- CPD&M integrates seamlessly with digital quality methods, leveraging real-time data analytics for continuous monitoring and adjustment of critical parameters.

Robust Design for Reliability (RDfR)

- CPD&M supports RDfR by identifying the parameters most crucial to product reliability, enabling the design of more resilient products.

The 12-Step CPD&M Process

- The 12-step CPD&M process will be explained in detail in Chapter 4.

Benefits and Challenges of CPD&M

Benefits

- Enhanced product reliability and performance.
- Reduced development and lifecycle costs.
- Improved understanding of product behavior.
- More effective risk management.
- Enhanced ability to meet regulatory requirements.

Challenges

- Requires significant resources and expertise.
- Can be time consuming, especially in complex systems.
- Necessitates a cultural shift toward data-driven decision-making.
- May require substantial investment in measurement and analysis tools.

CPD&M is a powerful and comprehensive approach to product development and lifecycle management. By meticulously identifying, analyzing, and managing critical parameters, organizations can create products that are not only high performing and reliable but also cost effective throughout their lifecycle.

The integration of CPD&M with other methodologies like PHM, digital quality implementation, and RDfR further enhances its effectiveness, providing a robust framework for engineering excellence.

As products become increasingly complex and operate in evermore demanding environments, the importance of CPD&M will only grow. Organizations that master this approach will be well positioned to lead in their respective industries, delivering products that consistently meet and exceed customer expectations while maintaining a competitive edge in the global marketplace.

References

1 Taguchi, G. and Yokoyama, Y. (1993). *Taguchi Methods: Design of Experiments.* Dearborn, Michigan: American Supplier Institute.

2 Readiness Plan, Transfer function-based design to improve product reliability and robustness in design for six sigma, Matthew Hu, Six Sigma Forum Magazine, American Society for Quality, August 2013.

3 Phadke, M.S. (1989). *Quality Engineering Using Robust Design.* Englewood Cliffs, NJ: Prentice-Hall.

4 Ambler, A.P. and Sheppard, J.W. (eds.) (2025). *Realizing Complex Integrated Systems.* Taylor & Francis Group (Chapter 8, authored by M. Hu & others).

5 Ushakov, I. (ed.) (1994). Handbook of Reliability Engineering.

6 Rao, C.R. (1992). Statistical inference and testing.

7 Pecht, M. (1990). Design for Reliability (DfR).

8 Physics-of-Failure: An Approach To Reliable Product Development by Michael Pecht and Abhijit Dasgupta, 1995, *95 IRW Final Report.*

9 Kimseng et al. (1999). Software reliability analysis.

10 Taguchi, G. (2004). *Subir Chowdhury and Yuin Wu.* Taguchi Quality Engineering Handbook: Wiley.

11 Clausing (2004). Reliability in Total Quality Management (TQM).

12 Lee, J., Wu, F., Zhao, W. et al. (2014). Prognostics and health management design for rotary machinery systems—reviews, methodology and applications. *Mechanical Systems and Signal Processing* 42 (1–2): 314–334.

13 A Survey of Health Indicators and Data-Driven Prognosis in Semiconductor Manufacturing Process, A. Thieullen, M. Ouladsine, J. Pinaton, *8th IFAC Symposium on Fault Detection,* August 29–31, 2012. Mexico City, Mexico.

14 Tsui, K.L., Chen, N., Zhou, Q. et al. (2015) Hindawi Publishing Corporation. Prognostics and health management: a review on data driven approaches. *Mathematical Problems in Engineering* 793161.

15 Box, G.E.P., Bisgaard, S., and Fung, C. (1988). An explanation and criticism of Taguchi's contribution to quality engineering. *Quality and Reliability Engineering, International* 4: 123–141.

16 Hu, M., & Yang, K. (2005) "Transfer function development in design for six sigma framework," Part I (SAE Technical Paper 2005-01-1215). SAE International.

17 Clausing, D. (1994). Total quality development. In: *A Step-By-Step Guide to World-Class Concurrent Engineering*. New York: ASME Press.
18 Biggio and Kastanis (2020). Machine learning reliability.
19 Kadry (2012). Reliability applications in diverse fields.

Further Reading

Akao, Y. (ed.) (1990). *Quality Function Deployment: Integration Customer Requirements into Product Design*. Cambridge MA: Productivity Press.

Clausing, D. and Frey, D.D. (eds.) (2006). *Improving System Reliability by Failure Mode Avoidance Including Four Concept Design Strategies*. John Wiley & Sons, Inc.

Davis, T.P. (2006). Science, engineering, and statistics. *Applied Stochastic Models in Business and Industry* 22: 401–430.

Deming, W.E. (1986). *Out of the Crisis*. Cambridge MA: Massachusetts Institute of Technology, Center for Advanced Engineering Study.

Deming, W.E. (1993). *The New Economics for Industry, Government, Education*. Cambridge MA: Massachusetts Institute of Technology, Center for Advanced Engineering Study.

Shewhart, W.A. (1931). *Economic Control of Quality of Manufactured Product*. New York: Van Nostrand.

Johannesson, P., Bergman, B., Svensson, T. et al. (2012). A robustness approach to reliability. *Quality and Reliability Engineering International* 29 (1): 17–32.

2

Robustness Thinking and Strategies for Reliability Development

CHAPTER MENU

2.1 Introduction

Reliability is one of the most important characteristics of an engineering system. Reliability can be interpreted as **robustness over time**, reflecting the system's ability to perform its intended functions consistently and accurately throughout its lifecycle, under a wide range of real-world conditions.

Traditional reliability approaches primarily evaluate reliability **after design decisions are largely fixed**, relying on testing, qualification, and field feedback to discover weaknesses. In contrast, **proactive reliability** is achieved by embedding robustness thinking directly into the design process, where design leverage is highest and corrective costs are lowest.

From this perspective, reliability is not viewed as a test result or a statistical afterthought, but as a **design attribute intentionally created during concept development, architecture definition, and early design optimization**. Robustness thinking provides the mechanism for this proactive reliability creation by explicitly addressing variation, uncertainty, and lack of knowledge before failures occur.

Experience shows that:

- Most field failures are **designed in**, not manufacturing accidents
- Late-stage fixes are costly and incomplete
- Reliability improvements are most effective **before hardware exists**

Accordingly, robustness thinking shifts the central reliability question from:

> *"Will this design survive testing?"*

to:

> *"Is failure physically possible given the design, variation, and real-world use?"*

This shift forms the foundation of proactive reliability.

2.1.1 Failure-Mode Avoidance: A Comprehensive Approach to Reliability

Failure-mode avoidance is a proactive design strategy that aims to prevent system failures by addressing both robustness issues and potential mistakes. This approach recognizes that reliability is a function of both the system's ability to withstand varying conditions (robustness) and the prevention of errors in design, manufacturing, and operation (mistake avoidance).

Understanding Failure Modes

1) **Soft Failures:**
   - Gradual degradation of performance
   - Often driven by sensitivity to noise factors
   - Strongly linked to insufficient robustness
2) **Hard Failures:**
   - Sudden or catastrophic loss of function
   - May result from robustness gaps, design mistakes, or compounded stresses

3) **Robustness thinking enables proactive reliability by:**
   - Reducing sensitivity to variation
   - Preventing stress amplification
   - Increasing failure distance between stress and strength

   Rather than discovering failure modes through testing, proactive reliability emphasizes identifying the physical conditions under which failure could occur and eliminating those conditions through design.

Robustness in Failure-Mode Avoidance

Robustness focuses on creating systems that can maintain functionality across a broad range of conditions. In the context of failure-mode avoidance:

1) **Preventing Soft Failures:**
   - Designing systems to adapt to varying environmental conditions
   - Implementing self-calibrating or self-adjusting mechanisms
   - Using materials and components that resist degradation
2) **Mitigating Hard Failures:**
   - Creating redundant systems for critical functions
   - Implementing fail-safe mechanisms
   - Designing for graceful degradation rather than catastrophic failure
3) **Strategies for Enhancing Robustness:**
   - Comprehensive environmental testing
   - Stress testing beyond normal operating conditions
   - Implementing adaptive control systems
   - Using simulation and modeling to predict behavior under various scenarios

Mistake Avoidance in Failure-Mode Avoidance

Mistake avoidance focuses on preventing errors that can lead to hard failures. This encompasses:

1) **Design Mistakes:**
   - Implementing rigorous design review processes
   - Using computer-aided design and simulation tools
   - Applying design for manufacturability (DFM) principles
2) **Manufacturing Mistakes:**
   - Implementing strict quality control measures
   - Using error-proofing techniques (poka-yoke)
   - Automating critical manufacturing processes
3) **Operational Mistakes:**
   - Designing intuitive user interfaces
   - Providing clear, comprehensive user manuals
   - Implementing safeguards against misuse
4) **Strategies for Mistake Avoidance:**
   - Conducting failure mode and effects analysis (FMEA)
   - Using checklists and standardized procedures
   - Fostering a culture of quality and attention to detail

5) **Strategies for Enhancing Robustness:**
   - Implementing Design for Six Sigma (DFSS) methodologies
   - Robust design
     - Concept design
     - Parameter design
     - Tolerance design
   - Robust technology development
     - Making technology insensitive to uncontrollable user conditions
     - Creating adaptive systems that can adjust to changing conditions
     - Implementing fault-tolerant designs with redundancy
     - Using advanced materials engineered for durability
     - Leveraging AI and machine learning for predictive maintenance

By focusing on both failure-mode avoidance and robustness, designers and engineers can create systems that not only perform well under ideal conditions but also maintain their reliability across a wide range of real-world scenarios and throughout their intended lifecycle. This comprehensive approach to reliability leads to products and systems that are more dependable, cost-effective in the long term, and capable of meeting or exceeding user expectations in diverse and challenging environments.

Failures come in various forms. Sometimes a component or part of the system breaks, leading to reduced or complete loss of functionality. In other instances, failures occur due to degeneration from wear or fatigue. Additionally, a product may fail to meet customer expectations if used in an unforeseen manner. From the customer's perspective, this is also considered a failure.

The material strength of a component or system can vary from unit to unit. Additionally, the demands placed on these units also differ, and prolonged exposure to harsh environments can lead to deterioration. Furthermore, there may be an insufficient understanding of the product's capacity, or the demands placed on it. If variation and lack of knowledge are potential root causes of failures, then an appropriate design criterion would be to make the product insensitive to these sources of uncertainty. This approach is known as robust design methodology. Robust design methodology systematically aims to create systems and component solutions that perform well despite noise factors, which are sources of variation that may compromise the system's functionality. It also takes into account other uncertainties—conditions about which we have incomplete information—and seeks to ensure robustness against these uncertainties.

Reliability performance is largely established during product development. Therefore, systematic efforts to prevent failures must be undertaken throughout the product development process, with approaches varying in different phases. The primary causes of poor product reliability are variation and lack of knowledge. This book presents a methodology for managing variation at different stages of product development. Additionally, it introduces a novel approach to address both lack of knowledge and variation within a unified framework.

The key aspects of robustness thinking in reliability design include:

1) **Variability:**

   Robustness thinking acknowledges that variations and uncertainties are inherent in real-world conditions. It involves anticipating and accommodating variations in materials, manufacturing processes, environmental factors, and usage conditions that could affect the performance of a system.

2) **Tolerance to External Factors:**
   Robust designs are tolerant to external factors such as temperature fluctuations, humidity, vibration, and other environmental conditions. This ensures that the system can operate reliably in a range of settings without a significant decrease in performance.
3) **Sensitivity Analysis:**
   Sensitivity analysis is a crucial component of robustness thinking in reliability design. It involves studying how variations or changes in inputs or parameters affect the system's output or performance. By identifying sensitive areas, designers can implement measures to mitigate the impact of variations.
4) **Redundancy and Fault Tolerance:**
   Robust designs often incorporate redundancy and fault-tolerant features. Redundancy involves having backup components or systems in place, ensuring that the failure of one part does not lead to overall system failure. Fault tolerance involves designing systems that can detect and recover from faults automatically.
5) **Testing Under Extreme Conditions:**
   Robustness thinking includes subjecting the system to testing under extreme or worst-case conditions. This helps identify weak points and vulnerabilities in the design. Testing under various scenarios ensures that the system can reliably perform even in challenging situations.
6) **Lifecycle Considerations:**
   Robustness is not just about initial reliability but also about maintaining performance over the entire lifecycle of the system. Considerations for maintenance, updates, and potential wear and tear are integrated into the design to ensure long-term reliability.
7) **Adaptability and Flexibility:**
   Robust systems are designed to adapt to changing conditions. This may involve incorporating flexible design features, adaptive control systems, or self-diagnostic capabilities. The ability to adapt enhances the overall reliability of the system in dynamic environments.
8) **Feedback Mechanisms:**
   Including feedback mechanisms in the design allows the system to continuously monitor its performance. This information can be used to make real-time adjustments, detect anomalies, and implement corrective actions, contributing to ongoing reliability.
9) **User Interaction and Error Tolerance:**
   Consideration of how users interact with the system is a part of robustness thinking. Design features that minimize the impact of user errors or misuse contribute to the overall reliability of the system.

Robustness thinking in reliability design involves anticipating and addressing variations, uncertainties, and challenges by incorporating features such as tolerance to external factors, redundancy, fault tolerance, adaptability, and ongoing monitoring. This approach aims to create systems that consistently perform well under a variety of conditions throughout their lifecycle.

A reliable product is insensitive to noise (uncontrollable user conditions) over time. In the context of product reliability, insensitivity to noise refers to a product's ability to maintain consistent performance and functionality despite uncontrollable or unpredictable conditions introduced by users or the environment. Noise, in this context, can be any form of interference, variability, or disturbance that the product might encounter during its use. Some key considerations related to the insensitivity to noise in the context of product reliability include:

1) **Robust Design:**
   - A reliable product is often designed to be robust, meaning it can withstand variations in operating conditions. Robustness ensures that the product is less susceptible to the effects of noise and continues to function as intended under a range of circumstances.
2) **Tolerance to Environmental Factors:**
   - Products may be exposed to various environmental conditions such as temperature fluctuations, humidity, and vibrations. A reliable product should be tolerant to these factors, minimizing the impact of environmental noise on its performance.
3) **User-friendly Design:**
   - Consideration of user behavior and potential misuse is crucial in ensuring product reliability. A reliable product is designed to tolerate variations in user behavior, reducing the likelihood of performance degradation due to unintended actions or usage patterns.
4) **Quality Control:**
   - Stringent quality control measures during manufacturing can contribute to product reliability. Ensuring consistency in the production process helps reduce variability between individual units of the product.
5) **Testing Under Real-World Conditions:**
   - Testing a product under real-world conditions, including exposure to noise and unpredictable factors, is essential for assessing its reliability. Simulating various scenarios during the testing phase helps identify potential weak points and areas for improvement.
6) **Fault Tolerance:**
   - Building fault-tolerant features into the product enhances its reliability. This may involve incorporating redundancy, error-checking mechanisms, and fail-safe modes to mitigate the impact of unexpected conditions.
7) **Adaptive Systems:**
   - Products with adaptive systems can dynamically adjust their behavior in response to changing conditions. Adaptive features contribute to insensitivity to noise by allowing the product to optimize its performance based on the current environment or user interactions.
8) **Regular Maintenance and Updates:**
   - Providing guidelines for regular maintenance and offering software or firmware updates can help maintain product reliability over time. This ensures that the product remains compatible with evolving technologies and user expectations.
   - Prognostic Health Management (PHM): Integrating PHM systems into the product can further enhance reliability by predicting potential failures before they occur.

PHM uses data analysis and machine learning to monitor the condition of components and systems in real-time, enabling proactive maintenance and minimizing downtime.

- Remote Monitoring: Utilizing remote monitoring tools can help track the performance and health of products in the field, facilitating timely updates and maintenance interventions based on actual usage patterns and conditions.

Insensitivity to noise is a desirable quality in a reliable product, and achieving this requires a combination of robust design, tolerance to environmental factors, user-friendly design, quality control, testing, fault tolerance, adaptive systems, and ongoing maintenance. By addressing these aspects, manufacturers can enhance a product's ability to consistently perform well over time, even in the face of uncontrollable user conditions.

Methods for the development of product robustness are essential steps to increase the quality and reliability in designs with reduced risks for their integration, manufacturing, and service. Robust design can enable engineering teams to prevent failures in later phases.

Use robust design as a problem prevention strategy, robust design contributes to reduced cycle times and related program costs, making them more predictable. Robust design, when implemented as a problem prevention strategy, can significantly contribute to the reduction of cycle times and related program costs in various industries. The essence of robust design is to create products, processes, or systems that are less sensitive to variation and external factors, thereby minimizing the likelihood of issues arising during development, production, or usage. Here's how robust design acts as a problem prevention strategy and leads to more predictable cycle times and program costs:

Early Identification and Prevention of Issues:

Robust design involves thorough analysis and testing during the early stages of product development or process design. By identifying potential issues and variations upfront, the team can implement preventive measures, reducing the likelihood of problems occurring later in the cycle.

Reduced Rework and Iterations:

A robust design aims to get things right the first time. By considering potential sources of variability and incorporating resilience into the design, the need for extensive rework or iterations is minimized. This not only accelerates the development cycle but also reduces associated costs.

Minimization of Design Changes:

Robust designs are less prone to frequent changes. By addressing potential issues and uncertainties proactively, the need for last-minute alterations is diminished. This stability in design contributes to more predictable schedules and lower program costs.

Improved Manufacturing Processes:

In manufacturing, robust design principles lead to processes that are less sensitive to variations in materials or environmental conditions. This results in higher production efficiency, fewer defects, and ultimately shorter cycle times.

Enhanced Supplier Relationships:

Robust design often involves collaboration with suppliers to ensure the quality and consistency of materials and components. Building strong relationships with reliable suppliers contributes to smoother production processes and can lead to more predictable costs and timelines.

Simulation and Modeling:

The use of simulation and modeling in robust design allows for virtual testing of various scenarios. This helps in understanding how the design will perform under different conditions, enabling the team to optimize the design before physical prototypes are built. This can significantly reduce the time required for testing and validation.

Customer Satisfaction and Fewer Warranty Claims:

A robust design that considers potential failure modes and addresses them proactively results in products that are more reliable. This, in turn, leads to higher customer satisfaction and fewer warranty claims, mitigating unexpected costs associated with product recalls or replacements.

Predictable Program Costs:

The combination of reduced rework, fewer design changes, and efficient manufacturing processes contributes to more predictable program costs. Teams can better estimate budget requirements and allocate resources with greater confidence.

Continuous Improvement:

Robust design is a mindset that encourages continuous improvement. By learning from each project, teams can refine their design and development processes, further enhancing their ability to prevent problems and reduce cycle times in future programs.

In summary, robust design, as a problem prevention strategy, contributes to the predictability of cycle times and program costs by addressing potential issues early in the development process, reducing the need for rework, minimizing design changes, improving manufacturing efficiency, and fostering a culture of continuous improvement.

2.2 What Is Robustness Thinking?

So, what exactly is robustness thinking? Let us consider the process of making popcorn the traditional way. What happens if the popcorn does not taste good or gets burned? The objective is to understand the reasons behind the failure to produce satisfactory popcorn and address those issues directly, such as the timing of the last pop, cooking duration, and temperature levels. Once an improved method is identified, we standardize it for consistent popcorn-making results. However, mastering this process may pose challenges, especially for individuals unfamiliar with popcorn-making techniques. In essence, these experiences are highly dependent on the user's circumstances.

In the popcorn story, the need of developing insensitivity to uncontrollable user environments, referred to as robustness, is essential to the objective of problem prevention in product designs. This approach precedes problem correction activities. Problem prevention yields downstream benefits in the development of new products. Products that are more robust transition into manufacturing and service more easily. The benefits of designs proven to be robust for a range of applications must include the knowledge being fed back upstream to the development of follow-on product designs. So, first, you develop robustness, and then you foster reliability.

Do things right—keep it consistent. This is the idea behind conventional problem solving such as Six Sigma. Is it simple? Six Sigma does not address the original design of the process or product; it merely improves on the existing product or process.

Contrastingly, let us look at today's popcorn bag. Making popcorn becomes much easier. The quality of cooked popcorn out of each bag today is much more consistent. Anyone including kids who want to make popcorn can put a bag of popcorn in a microwave and turn on a switch. There is no need to worry about the sound of the last pop and the temperature. In other words, the quality of popcorn is not sensitive to external factors, such as the popper's age and experiences. This is the idea of robust design which is a key element of DFSS and our approach in terms of robust design for reliability (RDfR).

DFSS is not simply a rehash of the lessons learned in Six Sigma, but a fundamentally different and proactive methodology. DFSS complements the Six Sigma improvement methodology but takes it one step further to address the design robustness. It seeks to overcome the traditional flaws of the product and the process during the design stage: conceiving and developing popcorn-making bag, rather than in the production quality control stage of preventing the burnt pot of popcorn.

"Do right things" is important for a company to stay in business and develop the right products to satisfy customer and, of course, make bottom line profits. Six Sigma focuses on low-hanging fruits (reducing costs and improving existing designs). DFSS concentrates its efforts on developing new and better designs for the engineer to exploit. DFSS is the "What" and "How To" of the engineering product development process. Its focus is to support the new product introduction process across the industry that:

- Translates the voice of the customer into an optimum design concept
- Quantifies and optimizes the design's performance and risk

A second example illustrates this well. If your company's business is producing and selling water faucets, a Six Sigma approach would investigate the warranty returns, and examine the process. This examination would result in the discovery that the quality of seals, ability to provide consistent water flow and temperature and water knobs were inadequate and replace them. A DFSS approach would come up with a design strategy to prevent the problem from happening.

The faucet in Figure 2.1 (adapted from [1]) has two design parameters (DP): a hot-water knob and a cold-water knob. When the hot-water knob is turned, temperature is affected but so is flow. Turning of the cold-water knob also affects temperature and flow. If a consumer

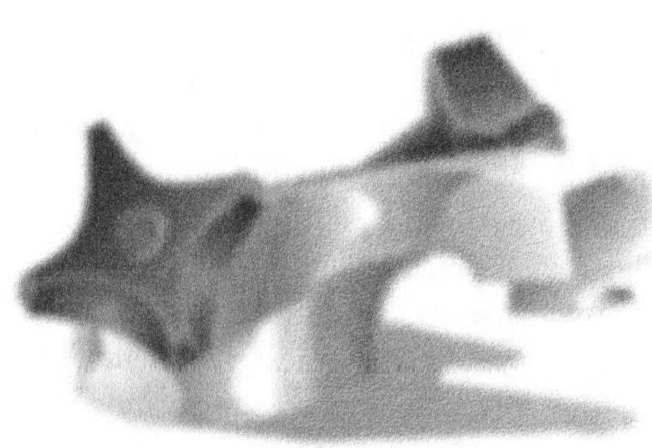

Figure 2.1 Review of Six Sigma and Design for Six Sigma (DFSS).

has optimized flow rate, then turns one of the knobs to optimize temperature, the flow rate is changed and is no longer optimal. Designs of this type can eventually satisfy customers only between the two designs parameters many times. Customers may run out of patience and are unsatisfied with this type of design.

However, a DFSS approach would take one step further and go beyond existing limited structure thinking—what can be done to improve a product functional performance and the ultimate in customer satisfaction? A team would reconfirm the voice of customers (VOC) and ensure proper translation of VOC into design functional requirements (FR). In the water faucet example, the FR of water faucet are:

FR1 = Control flow rate

FR2 = Control temperature

The DPs for Figure 2.1 are:

DP1 = Hot-water knob

DP2 = Cold-water knob

Obviously, both DP1 and DP2 in Figure 2.1 affect FR1 and FR2 at the same time. This type of design is known as coupled and the FR are not independent [1, 2] and, of course, the customer may experience inconsistent functional behavior and be unsatisfied. Coupling is about two functions affected by same one variable.

Consider the design in Figure 2.2 (adapted from [1]). This faucet has one handle, and the DPs are: lift the handle to adjust flow, and move the handle from side to side to adjust temperature. In this design, adjusting temperature does not affect flow, and adjusting flow does not affect temperature.

Figure 2.2 Robustness Thinking in DFSS Strategy.

Design 2.2 (Figure 2.2):

DP1 = Handle Lifting

DP2 = Handle Moving Side to Side

From an axiomatic design point of view, this design is superior because it maintains the independence of the FR. Of course, the design needs to be further optimized to desensitize the effects of user conditions (variations).

Just imagine while you are writing an important report and needs to be finished in 30 minutes, and you have to answer an important phone call at the same time. The function of writing and the function of answering phone calls are fighting for attention and conflict each other. Conflict causes coupling effects. Is it hard to concentrate? In reality, the designer often deals with more than two FR. Imagine what happens when a designer is working in a situation with a dozen or more FR. If the design is coupled, then optimization of one function may adversely impact several other functions. When these functions are fixed, the original function no longer works well. The designer is always tuning and band-aiding such a design, and the customer will never be completely happy. However, if the design is created in such a way that each of the FR is handled independently by the DPs, then each function of the design can easily be optimized. From a design structure standpoint, this kind of design is better and more robust due to the fact of minimum coupling effect. Taguchi's additive model (reduction of interaction effects by selecting proper system output response) can help us to obtain further information to reduce interaction effects [2, 3]. Interaction pertains to two design variables affecting the same function. Functional independence (Suh) or the interaction reduction (Taguchi's additive model) provides the opportunity to design and manufacture products that will attenuate all forms of unwanted variation, including variation due to conditions of customer use.

In order to be successful in today's business, any company needs to strategically plan all development projects with the right level and the right kind of development to achieve maximum efficiency. It has been estimated that 85% of the problems with new products not working as they should, taking too long to bring to market, or costing too much is the result of a poor design process (Ullman, 1997). A good design process is supported by a set of efficient methodologies. It has been widely accepted that the early phases of the engineering design process are the most critical to the technical and economic success of a new product. Therefore, the use of an efficient methodology for this crucial stage is most important. DFSS is such an efficient methodology—the What and How to engineer product development process to support a new product introduction process.

DFSS also effectively supports redesign and improvement of portions of a product (e.g., fix chronic problems). DFSS uses statistical tools to understand, optimize, and control key factors that deliver critical customer attributes robustly in the presence of noise. In quantifying risk, DFSS promotes business discussions regarding product delivery quality early in the product development process, while the design can still benefit from the VOC insights.

Traditionally companies accepted three or four sigma performance levels as the norm, even though these created between 6,200 and 67,000 problems per million opportunities. The Six Sigma standard of 3.4 problems per million opportunities is a response to the increasing expectations of customers and the increased complexity of modern products and processes. Most companies have not taken the time or made the effort that is required to:

- Learn what their customers really want.
- Develop critical-to-satisfaction metrics in terms of design variables and the scorecard for risk assessments.
- Assess field and design robustness for testing strategy development.

DFSS requires applying resources to finding out what the customers really want, and then devoting the entire project to meeting the needs and desires of these customers. The result of DFSS implementation is higher profits, increased customer satisfaction, and high levels of quality and reliability.

The need for DFSS succeeds by measuring defective processes, analyzing the causes of the defectives and then improving them. "Design for Six Sigma" is more proactive by starting with the VOC and then doing the right thing. DFSS focuses on delivering the best product and developing the best process at the same time.

Robustness thinking in reliability development involves creating systems, products, and processes that can perform effectively and consistently under variation, uncertainty, and disturbance. These variations arise from manufacturing, environment, customer usage, aging, and interactions between subsystems.

A key premise of robustness thinking is that **variation and lack of knowledge are the primary root causes of poor reliability**. Because many of these sources cannot be eliminated, the appropriate design objective is to make the system **insensitive to them**.

From a proactive reliability standpoint, robustness thinking:

- Prevents failure modes rather than reacting to them
- Treats reliability as a **leading design indicator**, not a lagging metric
- Focuses on variability of functional response rather than failure counts

A reliable product, therefore, is one that maintains its functional performance despite uncontrollable user environments and real-world uncertainties.

2.3 The Challenge and Limitation of Conventional Reliability Approach

The comparison between reliability and robustness can manifest in various ways. In this section, our focus is on definition, metrics, and the ability to proactively design from a quality and reliability perspective. Both reliability engineering and robust engineering boast robust metrics, and understanding when and how to apply each provides a significant advantage to a product engineering team. The key lies in appreciating the advantages of each approach and combining their synergies to achieve highly reliable products, ensuring better customer satisfaction at the lowest possible cost and within a shorter development cycle time.

We discuss how to think about reliability in terms of uncertainty, that is, variation and lack of knowledge. Conventional reliability is defined as the probability of a product performing its intended function for a specified life under the encountered operating conditions.

Variation and uncertainty have become important parts of our current world view. In an industrial setting, concepts like Six Sigma and DFSS have become important.

2.3.1 Uncertainty—Variation and Lack of Knowledge

We want to emphasize the importance of understanding variation in creating reliable products. In addition, we will broaden our perspective to discuss other uncertainties that may arise when assessing product reliability. The first type of uncertainty stems from random variation, while the second type arises from our lack of knowledge, such as when modeling product characteristics or estimating parameters.

Sources of Variation: The well-known bathtub curve in reliability discourse can be interpreted through an understanding of variation. The corresponding sources of variation affecting the product are referred to as noise factors. Noise factors are the primary reason for variability in design demands and capacities, and they are often the cause of many design failures. These factors are typically expensive or even impossible to control under operating conditions, so their potential influence must be managed through other means. There are two types of noise factors: those that create a demand on the product and those that affect its capacity to meet that demand. Major noise factors on the demand side include:

- Customer usage, which encompasses the different ways customers operate the product.
- External environmental conditions, including factors like temperature, humidity, wind speed, and road conditions.

- Internal environmental conditions, involving interactions between neighboring subsystems, such as transmitted heat or vibrations.
- Unit-to-unit variation due to production conditions, such as:
  - Material differences
  - Variations in the manufacturing process
- Changes over time due to field operation, referring to long-term deterioration resulting from the product's usage and its environment.

Unit-to-unit variation, after extensive improvement efforts, can typically be characterized as originating from a process that is predictable within certain limits. A probabilistic model may suffice, especially if thorough measures have been taken to reduce assignable causes of variation. However, the demand side presents greater complexity.

Generally, customer-to-customer variation is less predictable than manufacturing processes. This unpredictability extends to the usage environment (as a demand) and its long-term effects on product capacity. Once a specific customer starts using a product, that unit's reliability is largely determined by that particular customer's demands. Consequently, product reliability becomes an average of each customer's unit reliability.

This means that overall product reliability can be high, even if a specific customer's unit fails to meet their particular demands or expectations. In this context, using reliability as a design criterion becomes challenging. Therefore, we propose adopting a robust design methodology and its associated design criteria as a more effective approach to managing reliability in the product development process.

Uncertainty stemming from knowledge gaps: When forecasting product properties, we encounter not only random variations but also other uncertainty sources, including model errors, parameter uncertainties, and human mistakes. There are several ways to categorize these uncertainties (as discussed by Melchers, Ditlevsen & Madsen, and Lodeby).

One primary distinction is between aleatory and epistemic uncertainties:

- Aleatory uncertainties: These are inherent, underlying uncertainties. Examples include the variability in product lifespan and the range of load conditions across different customers.
- Epistemic uncertainties: These uncertainties arise from knowledge limitations. They can be mitigated through:
  - Gathering additional data or information
  - Improving modeling techniques
  - Enhancing parameter estimation methods

Understanding this distinction is crucial, as it helps identify which uncertainties are inherent to the system and which can be reduced through improved knowledge and methods.

Melchers provides a comprehensive classification of various uncertainty types. In our reliability modeling approach, we will concentrate on three specific categories of uncertainties, which are also highlighted by Ditlevsen & Madsen. This simplified focus allows us to streamline our reliability modeling while still addressing the most crucial aspects of uncertainty as identified by key researchers in the field.

2.3.1.1 Random Variation or Physical Uncertainty

This refers to the inherent randomness in a phenomenon. For example, the strength variation between different components. It's sometimes called scatter, randomness, or noise. Importantly, a noise factor is a source of this random variation.

2.3.1.2 Statistical Uncertainty

This arises from statistical estimation of model parameters based on available data. For instance, the uncertainty in estimating parameters in a regression model that describes life as a function of load level. It occurs because:

- Observations may not perfectly represent the variable, potentially leading to data bias.
- Different sample data sets typically produce varying statistical estimates.

2.3.1.3 Model Uncertainty

This is associated with using simplified relationships to represent complex real-world phenomena. For example, a finite element model used to calculate stresses is only an approximation of the actual stress state. Model uncertainty often stems from knowledge gaps and can be reduced through: further research or increased data availability.

Understanding these three types of uncertainties is crucial for comprehensive reliability modeling, as they capture different aspects of variability and imperfect knowledge in engineering systems.

2.3.1.4 Among These Three Types of Uncertainties

- Random variation clearly falls under the category of aleatory uncertainty. This type of uncertainty is inherent and irreducible in nature.
- Statistical uncertainty and model uncertainty, on the other hand, should be classified as epistemic uncertainties. These types of uncertainties stem from limitations in our knowledge or understanding.

The key distinction lies in their potential for reduction:

- Aleatory uncertainty (random variation) is inherent and cannot be reduced.
- Epistemic uncertainties (statistical and model uncertainties) can be minimized through improved knowledge, more accurate data collection, advanced modeling techniques, or enhanced understanding of the system in question.

This classification helps in determining which aspects of uncertainty we can work to reduce and which we must account for as unavoidable variability in our systems and models.

Another significant type of uncertainty arises from human factors. These uncertainties stem from human involvement in design, system operation, usage, and other areas, and are primarily due to human errors and interventions. Unlike other types of uncertainties, these cannot be controlled solely through statistical methods.

Failures caused by misuse, gross errors, and human mistakes are mainly addressed through quality management practices. However, human-related uncertainties can also be mitigated by:

- Implementing system changes to enhance robustness (as discussed in Section 2.3).
- Applying additional safety factors.

By addressing human factors through these methods, we can better manage the uncertainties they introduce and improve overall system reliability.

2.3.2 Traditional Reliability Challenges

Traditional reliability engineering faces several significant challenges, particularly when quality data is unavailable or insufficient, especially in the early design phase. Here are some of the key issues:

1) **Determining Density Functions:**
   - The challenge lies in accurately determining the density functions for reliability predictions when there is a lack of quality data. This is particularly problematic during the early stages of design when empirical data is scarce.
2) **Reliability Growth and Improvement:**
   - Reliability growth or improvement efforts are centered around identifying and addressing issues. However, without sufficient data, it becomes difficult to pinpoint the exact areas that need improvement.
3) **Reliability as a Lagging Indicator:**
   - Reliability in the market, or as predicted during development, often serves as a lagging indicator. This means that by the time reliability issues are identified in the field, it may be too late to make cost-effective corrections.
4) **Data Collection and Censoring:**
   - One common difficulty with reliability data is censoring, where the observation period ends before all units have failed, leaving some as survivors. This can lead to a lack of failures in the data, making it harder to prove the reliability of a product.
5) **Model Errors and Parameter Uncertainties:**
   - Traditional reliability prediction models often suffer from model errors and parameter uncertainties. These issues arise because the models may not accurately reflect real-world conditions or because the parameters used in the models are based on limited or outdated data.
6) **Human Factors:**
   - Uncertainties due to human factors, such as errors and interventions during design and operation, are another significant challenge. These uncertainties are not easily controlled through statistical methods and require robust quality management practices and system changes to mitigate.
7) **Outdated Reliability Prediction Methods:**
   - Many traditional reliability prediction methods rely on outdated data and models. For example, some military standards have not been updated for decades, leading to inaccurate predictions and potentially costly design decisions.
8) **Complexity of Modern Systems:**
   - The multi-disciplinary nature of modern systems, which include hardware, software, and mechanical components, demands a comprehensive approach to reliability. Traditional methods may not adequately address the interactions and dependencies between these different components.

As depicted in Figure 2.3, calculating the probability of success necessitates information about the mean, standard deviation, and f(.), where f(.) represents the probability density function.

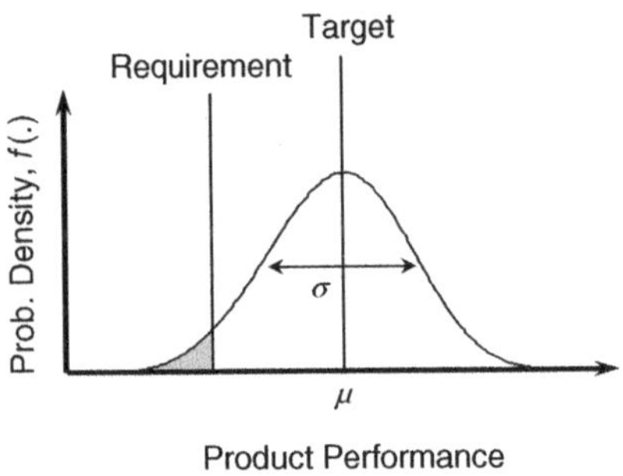

Figure 2.3 Product Performance and Reliability.

The challenge lies in determining these density functions when quality data is unavailable or insufficient, particularly in the early design phase. Reliability growth or improvement is centered around addressing issues, with reliability in the market or as predicted during development serving as a lagging indicator. Addressing these challenges requires a combination of improved data collection methods, updated models, robust design methodologies, and a focus on both aleatory and epistemic uncertainties. By doing so, reliability engineers can better predict and enhance the reliability of products throughout their lifecycle.

Robustness is a critical concept in product design and engineering that refers to a product's ability to consistently perform its intended function despite variations in its operating environment or usage conditions. Let us expand on this concept and its implications:

1) **Definition and Significance:**
   Robustness is the ability of a product to consistently perform its intended function in the presence of an uncontrollable user environment (noise) throughout its intended life. In other words, the product is insensitive to noise. This characteristic is crucial for ensuring product reliability and customer satisfaction across diverse usage scenarios.
2) **Evaluation Metrics:**
   As shown in Figure 2.4, evaluating robustness requires knowledge of the mean and standard deviation without the need to know the density function. This approach simplifies the assessment process, allowing engineers to focus on key statistical measures rather than complex probability distributions.
3) **Relationship to Reliability:**
   Robustness serves as an engineering metric for reliability. While reliability typically focuses on the probability of failure over time, robustness addresses the product's

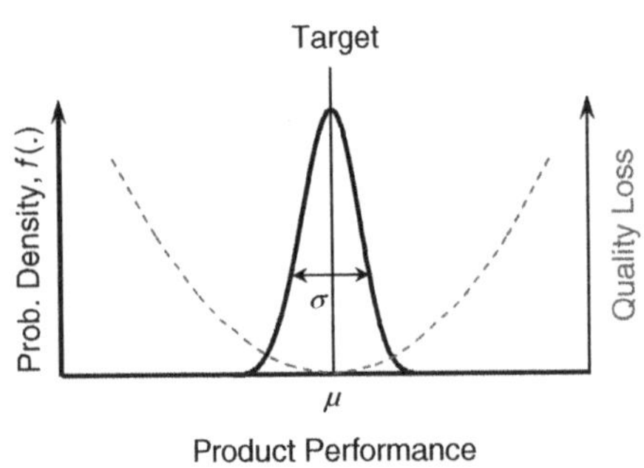

Figure 2.4 Product Performance and Robustness.

performance consistency across various conditions. A robust product is more likely to be reliable, as it can withstand environmental variations without significant performance degradation.

4) **Quality Improvement:**

   Improving robustness reduces quality loss. By making a product less sensitive to noise factors, manufacturers can minimize variations in performance and reduce the likelihood of product failures or customer dissatisfaction.

5) **Manufacturing Implications:**

   In manufacturing, robustness enhances the process capability index. This index measures how well a process can produce output within specified limits. A more robust process is capable of consistently producing products that meet specifications, even with minor variations in input materials or manufacturing conditions.

6) **Robust Design for Reliability (RDfR):**

   Robustness is one of the crucial efforts in RDfR in manufacturing. RDfR is an approach that aims to create designs that are inherently less sensitive to variations in manufacturing, environment, and usage.

   This methodology helps in:

   - Reducing the need for tight tolerances.
   - Minimizing the impact of uncontrollable factors.
   - Improving overall product quality and reliability.

7) **Applications:**

   Robustness is particularly important in industries where products are exposed to varying conditions, such as:

   - Automotive: vehicles must perform consistently in different climates and road conditions.
   - Electronics: devices should function reliably despite variations in power supply or user handling.
   - Aerospace: components must withstand extreme temperature and pressure changes.

8) **Design Strategies:**

   To improve robustness, engineers may employ strategies such as:

   - Parameter design: optimizing product parameters to minimize sensitivity to noise factors.
   - Tolerance design: determining appropriate tolerances for critical components.
   - Redundancy: incorporating backup systems for critical functions.

9) **Economic Impact:**

   Robust products typically have lower warranty costs, higher customer satisfaction, and better brand reputation. This can lead to increased market share and profitability for manufacturers.

By focusing on consistent performance overtime across varying conditions, robust design practices help create products that are more reliable, have higher quality, and provide better customer satisfaction. This approach is increasingly important in today's complex and globalized manufacturing environment.

A condensed comparison between robust design and reliability is presented in Table 2.1 below. For the purposes of this discussion, additional crucial information, such as testing strategies, is not covered in this table.

Table 2.1 Robust Design and Reliability Comparison.

<table>
<tr><th></th><th>Robust Design</th><th>Reliability</th></tr>
<tr><td>Definition</td><td>Robust Design focuses on creating products or processes that can consistently perform their intended function despite variations in the operating environment or usage conditions. It aims to transform input to output as closely to the ideal function as possible, even in the presence of uncontrollable user environment factors (noise)</td><td>Reliability is defined as the ability of a system to perform its required functions under given conditions at a specified time. It focuses on the consistency of performance over time</td></tr>
<tr><td>Focus</td><td>The primary focus is on maximizing design intent by developing functional robustness in three key areas:
Technology: Selecting and developing robust technologies that are inherently less sensitive to variations
Concept Design: Creating design concepts that are fundamentally more robust
Detail Design: Optimizing parameters and tolerances to enhance robustness</td><td>Reliability efforts concentrate on increasing the system's ability to function without failure over time. This is typically achieved through a cumulative implementation of "build–test–analyze–fix" cycles</td></tr>
<tr><td>Approach</td><td>The core approach in Robust Design is to make the design insensitive to uncontrollable user environment (noise) through:
• Robust technology selection and development
• Concept design optimization
• Parameter design
• Tolerance design</td><td>The primary approach in reliability engineering is reliability growth—an iterative process that involves:
• Testing the system
• Identifying problems
• Analyzing causes of failures
• Fixing identified problems
• Verifying solutions
• Repeating the cycle to continuously improve reliability</td></tr>
<tr><td>Metrics</td><td>Signal-to-Noise ratio (S/N): This is a key metric used to measure intended design output vs. unintended design output. A higher S/N indicates that it's easier to separate the signal (e.g., energy, information, and material) from the unintended output
S/N is considered an engineering measure of reliability and a leading indicator, as it can predict potential reliability issues before they occur in the field</td><td>Reliability uses several lagging indicators, including:
• Failure rate: The frequency with which a system or component fails
• Mean Time Between Failures (MTBF): Average time between system failures
• Mean Time to Failure (MTTF): Average time until a nonrepairable system fails
• Mean Time to Repair (MTTR): Average time required to repair a failed system
• Yield: Percentage of units that pass quality control
• Process Capability Index (Cp/Cpk): Measures how well a process meets specification limits</td></tr>
</table>

Key Differences and Relationships

1) **Proactive vs. Reactive:** Robust Design is more proactive, aiming to prevent issues before they occur, while traditional reliability approaches are often more reactive, addressing issues as they are discovered through testing and field use.
2) **Leading vs. Lagging Indicators:** Robust Design metrics like S/N ratio are leading indicators, potentially predicting reliability issues before they manifest. Reliability metrics are typically lagging indicators, measuring performance after the fact.
3) **Design Phase Focus:** Robust Design is heavily focused on the early design phases, aiming to build in quality from the start. Reliability efforts often continue throughout the product lifecycle, including after production begins.
4) **Scope:** Robust Design considers a broader range of potential variations and noise factors, while reliability often focuses more narrowly on preventing specific failure modes.
5) **Complementary Nature:** While distinct, these approaches are complementary. A robustly designed product is likely to be more reliable, and reliability testing can inform robust design practices in future iterations.

The development of robustness in a product's design focuses on preventing problems and optimizing robustness at the integrated system level. Robustness metrics, which are leading indicators, are crucial during the development phase. Achieving robustness early in development enables shorter cycle times in later phases, as it reduces the need for extensive physical testing and traditional test–fix–test cycles. Addressing issues late in the process can lead to delays in market entry and increased manufacturing costs due to tool modifications and scrapped inventories.

By integrating both robust design principles and traditional reliability approaches, organizations can create products that are less sensitive to variations and more consistently reliable over time.

Early robustness development in manufacturing can diminish process variability, yielding valuable benefits for manufacturing outputs, cycle time, and costs. This proactive approach centers on designing processes that are less sensitive to variations and uncertainties. By addressing potential challenges at the outset, this strategy can yield numerous benefits, including improved manufacturing yields, reduced cycle times, and lower costs. Here's how early robustness development contributes to these advantages:

Reduced Variability:
Robust design principles applied early in the manufacturing process aim to identify and mitigate sources of variability. This leads to processes that are more stable and less prone to fluctuations in inputs or environmental conditions. Reduced variability contributes to higher consistency in product quality and performance.

Improved Manufacturing Yields:
A manufacturing process designed with robustness in mind is less likely to produce defects or substandard products. This improvement in process stability and reliability leads to higher manufacturing yields, meaning a greater percentage of produced units meet quality standards, reducing waste and rework.

Faster Problem Resolution:
Early robustness development involves comprehensive testing and analysis during the design phase. This proactive approach allows teams to identify potential issues before

mass production begins. As a result, any necessary adjustments or improvements can be made early, minimizing disruptions and delays in the manufacturing process.

Shortened Cycle Times:
Processes designed for robustness are often more efficient and predictable. With fewer disruptions and a lower likelihood of unexpected issues, the entire manufacturing cycle can be streamlined. Shortened cycle times result in quicker turnaround from production initiation to product delivery.

Cost Reduction:
Early identification and mitigation of variability reduce the need for extensive rework, repairs, or scrap. This, in turn, lowers operational costs associated with addressing defects and waste. Additionally, fewer disruptions in the manufacturing process contribute to cost savings by avoiding unexpected expenses.

Optimized Resource Utilization:
Robust processes are more reliable and less prone to unexpected breakdowns or interruptions. This reliability allows for better planning and utilization of resources such as materials, equipment, and labor. Optimized resource allocation contributes to cost-effectiveness.

Enhanced Product Quality:
Robust design principles ensure that the manufacturing process consistently produces products that meet or exceed quality standards. Improved product quality not only satisfies customer expectations but also reduces the need for costly post-production quality control measures.

Consistent Performance:
Processes designed with robustness in mind tend to exhibit consistent performance over time. This consistency is crucial for meeting production targets, maintaining customer satisfaction, and establishing a reliable reputation in the market.

Early Return on Investment (ROI):
Investments made in early robustness development pay off by avoiding costly setbacks and delays. The sooner a reliable and efficient manufacturing process is established, the earlier the return on investment can be realized.

In summary, early robustness development in manufacturing has far-reaching benefits, including reduced variability, improved yields, shortened cycle times, and lower costs. This proactive approach sets the foundation for efficient and reliable production processes, contributing to overall competitiveness and success in the marketplace.

The impact of product development on revenue streams and reliability is substantial. It is both cost-effective and time-efficient to design the product to be insensitive to uncontrollable user environments during the upfront design phase. The incorporation of robustness development in the DFR process yields advantages by reducing the need for early physical testing and traditional test–fix–test cycles. Achieving robustness in the early stages of development facilitates shorter cycle times in the subsequent design phases.

Every design encounter uncertainty and natural variation. Questions about usage, response to environmental factors, or adaptations to changes in manufacturing and operational processes add complexity. Designing reliable and robust systems that consistently perform as expected becomes particularly challenging when uncertainties arise from alterations in manufacturing, operational, or environmental conditions.

Do customers consistently use our products under optimal conditions? Is our manufacturing process always conducted under optimal conditions? The answer to both these questions is no. The challenge lies in the fact that variation is ubiquitous and inherent in all products and processes. Products or processes sensitive to the effects of variation significantly contribute to issues in quality, reliability, and, ultimately, customer dissatisfaction. How can we make our products or processes insensitive to this variation?

2.3.3 Demand and Capacity–Statistical Modeling

Let us say that we are interested in the life of a product in service, where we may think of the product as a single component or as a system consisting of a number of components. The definition of the failure of the product may differ between different applications: the failure of any single component, the failure of a critical part, the lack of function of an important subsystem, or the lack of function of the whole system. An overall metric of reliability is the observed failure frequencies in service. This metric is directly coupled to our main interest, the life of a product in service.

Failures resulting from poor designs, harsh usage, and subpar quality can be attributed to a lack of robustness. These issues are well suited for systematic engineering approaches, where physical and statistical modeling serve as powerful tools for continuous improvement. In this context, we propose using the demand-and-capacity concept as our primary statistical tools. To implement this, we first need to adopt a statistical perspective.

The statistical approach to reliability considers the product's strength as its capacity, modeled as a random variable and examined across the population of products. However, a product's lifespan is also influenced by its usage during service. This means that the load, representing the demand, varies over time and among different users in many applications. Consequently, the statistical approach also analyzes the population of usage patterns. As a result, the statistical distribution of a product's life is a combination of both the strength (capacity) and load (demand) populations.

This perspective allows for a more comprehensive understanding of product reliability, taking into account both inherent product characteristics and real-world usage conditions. By considering these two aspects together, we can better predict and improve product performance and longevity across various user scenarios.

The product's lifespan is frequently analyzed using a diagram that illustrates the interaction between load and strength, commonly referred to as the load–strength diagram or the demand–capacity diagram (Figure 2.5). This diagram presents two statistical distributions, depicted as both relative histograms and their hypothetical probability density functions.

Strength Distribution: For engineering products, strength is influenced by various factors beyond just material properties. These include geometry, connections, surface irregularities, heat treatments, and post-treatments such as blasting or shot peening. When assessing reliability, the strength distribution should encompass all these essential sources of variability.

Geometry is defined by drawings and may include classifications, such as weld geometry. The variability in these factors is typically constrained within specified limits. However, the distribution within these limits is heavily influenced by manufacturing control, and the limits themselves can be somewhat ambiguous due to measurement uncertainty.

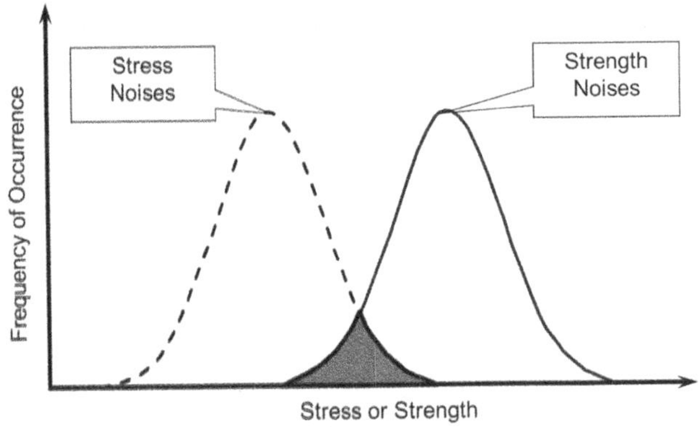

Figure 2.5 Product Performance and Robustness.

These factors can make assumptions about the distribution type rather arbitrary. When such influences significantly impact variability, it is important to exercise caution to avoid subjectivity.

Surface roughness, heat treatments, and post-treatments can significantly alter the material's strength properties. These effects are often unevenly distributed across the product's surface or volume. When combined with stress concentrations at geometric irregularities, they create hot spots. These hot spots are likely few in number, which can cause the weakest link theory to fail in accurately predicting the statistical distribution type. Consequently, assumptions about the statistical properties of these uncertainties are inherently weak, making any predictions about the distribution type questionable.

Load Distribution: The load distribution reflects the range of ways a product is used. This requires considering variations and uncertainties in factors such as the operating environment, human behavior, mission profiles, and extreme events.

By studying customer behavior, it may be possible to capture the overall variability related to human behavior, environment, and mission profiles. However, rare extreme events, such as a car hitting a rock on the road or a human error overriding expected behavior, are unlikely to be identified through customer surveys. Consequently, the tails of the load distribution, which represent these rare events, are crucial but typically cannot be accurately estimated due to the small sample sizes available in reliability engineering. We recommend treating extreme events separately from the statistical load–strength analysis. The statistical analysis should be confined to properties that can be reasonably measured and assigned statistical values. Any assumptions or subjective decisions must be transparent and open to scrutiny, rather than being obscured within a mathematical model.

This visual representation allows for a clear comparison between the product's capacity to withstand stress (strength) and the actual stress it experiences during use (load). By examining the overlap and relationship between these two distributions, engineers can gain valuable insights into the product's reliability and potential failure modes.

Regarding the Choice of Distribution Type: In reliability assessments of complex structures that include significant variability sources beyond material properties, there is often a lack of physical justification for selecting any specific distribution type. As a result, knowledge of statistical properties beyond mean values and standard deviations is typically limited. Consequently, methods that rely on full statistical distributions, such as probabilistic software integration or Monte Carlo simulations, are often highly sensitive to subjective choices of input distributions. This sensitivity can significantly impact the reliability assessment's accuracy and validity.

2.3.4 Deterministic vs. Probabilistic Design

In a typical design approach, the designer considers the likely extreme values of load and strength and ensures that an adequate safety factor is provided.

Strength: This refers to the capacity of a product, subsystem, or component to endure stress, and is considered a random variable. Stress can result from various loads, including electrical, thermal, mechanical, or chemical. Strength encompasses any resisting physical property, such as hardness, tensile strength, melting point, or adhesion. For example, a transistor gate in an integrated circuit fails when the applied voltage causes a local current density and subsequent temperature rise that exceeds the melting point of the conductor material.

Safety Factor (SF): This term describes the structural capacity of a system beyond the expected or actual loads. Essentially, it measures how much stronger the system is than it typically needs to be for an intended load.

Random variation and other uncertainties in both demand and capacity lead to the use of safety factors or safety distances. This is particularly crucial for failure modes caused by long-term deterioration and unforeseen events resulting from poor manufacturing or misuse.

Traditionally, safety factors have been based primarily on experience and set at rather arbitrary levels, without rational justification. Conversely, there is a trend toward using advanced probabilistic methods to determine safety factors that ensure well-defined, low probabilities of failure. Both approaches rely heavily on subjective judgments: the former due to the absence of a rational method and the latter due to insufficient knowledge about the statistical properties of noise factors and other sources of uncertainty.

A balanced approach between deterministic and fully probabilistic methods involves using statistical techniques on the available information. This knowledge is typically limited to nominal values and basic measures of uncertainty around these values. While this limited information makes advanced probabilistic and Monte Carlo simulations impractical, it still allows for more rational decision-making than traditional deterministic methods.

A deterministic approach to engineering analysis, which relies on fixed values and safety factors, can lead to inconsistent products due to over-design or under-design. This inconsistency arises from limitations in the available information about design, service conditions, material properties, and manufacturing parameters inherent in the deterministic approach.

Under a basic stress–strength framework, DPs are selected to ensure that the "strength" of the product exceeds the "stress" it will encounter. Figure 2.6 illustrates this simple stress–strength analysis. Recognizing that both stresses and strengths can vary, safety margins are established to minimize the risk of failure, typically based on:

- Rules of Thumb
- Past Experience

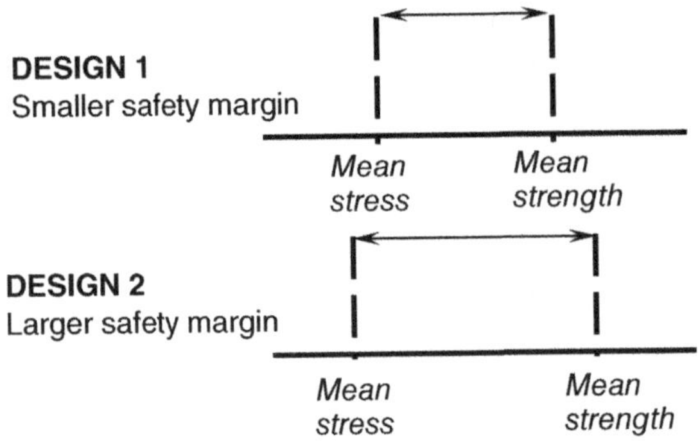

Figure 2.6 Deterministic.

A common question might be: which design is better, Design 1 or Design 2? The answer could be Design 2, as it has a larger safety margin.

This traditional approach may become inadequate in the future due to several factors:

- Increasing customer expectations for reliability.
- The need for improved energy efficiency.
- Growing global market competition in terms of cost and time to market.

Without incorporating probabilistic analysis, our understanding of how variability affects performance is limited. Probabilistic analysis helps illuminate the shape and dispersion of variability caused by noise factors. The area where stress and strength distributions overlap, known as the interference region, defines the probability of failure and thus determines reliability. Importantly, a design with a larger safety factor may not always result in higher reliability. The actual reliability depends on the variability of both stress and strength distributions, as illustrated in Figure 2.7. This concept challenges the conventional wisdom that a larger safety margin always leads to better reliability. By adopting a probabilistic approach, designers can:

- Better understand the impact of variability on product performance.

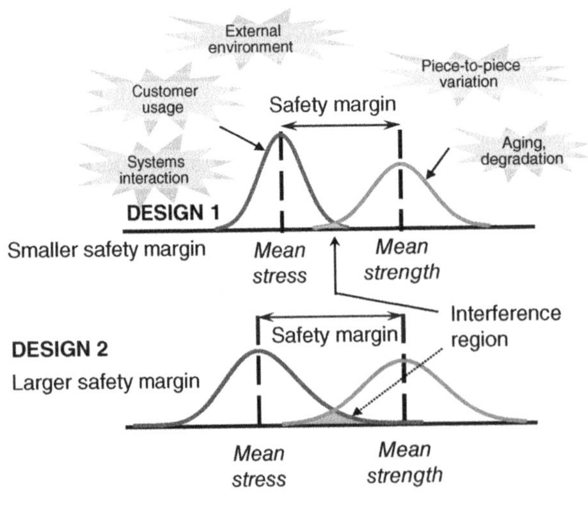

Figure 2.7 Probabilistic Design.

- Make more informed decisions about design trade-offs.
- Optimize designs for both reliability and efficiency.
- Potentially reduce over-engineering while maintaining or improving reliability.

This shift toward probabilistic design methods allows for a more nuanced and effective approach to product development, balancing reliability, cost, and performance in an increasingly competitive global market.

The mean strength of a product can decrease due to various factors in the user environment, including:

- Manufacturing variations from piece to piece.
- Aging and deterioration over time.

Conversely, the mean stress experienced by a product can increase due to uncontrollable environmental factors, such as:

- External environmental conditions.
- Variations in customer usage patterns.
- Interactions within the system or with other systems.

These shifts in mean strength and stress highlight the dynamic nature of product reliability in real-world conditions. They emphasize the importance of considering not just initial DPs, but also how these parameters may change over the product's lifecycle and under different usage scenarios. Understanding these potential changes is crucial for designing products that maintain their reliability throughout their intended service life.

In adopting a probabilistic approach to design, designers shift their perspective from viewing variables as fixed, single values to considering them as probability distributions. This probabilistic design methodology allows for the prediction of how variability propagates through a system. By understanding this flow of randomness, designers can:

1) Anticipate the impact of variability on system performance.
2) Make informed adjustments to reduce the effects of random variations.
3) Improve overall product quality and reliability.

This approach offers several advantages:

- It provides a more realistic representation of real-world conditions, where variables often fluctuate.
- It allows for better risk assessment and management in the design process.
- It enables designers to identify and focus on the most critical variables affecting system performance.
- It facilitates the creation of more robust designs that can withstand a range of operating conditions.

By embracing this probabilistic mindset, designers can create products that are not only optimized for ideal conditions but are also resilient to the inherent variability present in materials, manufacturing processes, and usage environments. This leads to more reliable, higher-quality products that better meet customer expectations across a wide range of scenarios.

The probabilistic approach, when combined with robustness thinking, motivates engineers to go beyond traditional design methods. It encourages them to seek out strategies like robust design to optimize product quality. The primary goal of this approach is to create products whose performance remains consistent despite variations in:

1) Manufacturing processes.
2) Material properties.
3) Operational conditions.
4) Environmental factors.

As demonstrated in a gerotor pump design, Dr. Taguchi's parameter design methodology aims to optimize a gear set by maximizing the safety factor. By applying this parameter design to gerotor pump design, engineers can develop more robust and efficient systems that are less sensitive to variations in manufacturing and operating conditions, ultimately enhancing performance and reliability.

Importantly, this optimization aims to achieve insensitivity to these variations while minimizing costs. This approach offers several benefits:

- It leads to more reliable products that perform consistently across a range of conditions.
- It can reduce manufacturing and warranty costs by minimizing the impact of variations.
- It encourages innovative thinking in design, focusing on creating inherently robust systems.
- It can improve customer satisfaction by delivering products that maintain their performance over time and in various usage scenarios.

By adopting this mindset, engineers are challenged to:

- Identify key sources of variation that affect product performance.
- Develop design solutions that minimize the impact of these variations.
- Balance performance, robustness, and cost considerations.
- Implement strategies that improve product quality without significantly increasing production costs.

This approach represents a shift from merely meeting specifications to creating designs that are inherently resistant to variability, resulting in higher-quality products that are more competitive in the market.

Categorically, there are five strategies for improving robustness and thus reliability:

- Change the design concept or technology.
- Make the design insensitive to noise factors.
- Reduce or remove the noise factors.
- Use a compensation device (e.g., dynamically tuned absorbers).
- Send the failure mode to another part of the system (trade-off) where it will do less harm.

As we indicated in Introduction, the second strategy for making the design insensitive to noise factors is the focus of this tutorial.

Phadke [4] states that there are three fundamental ways of improving the reliability of a product during the design stage:

- Reduce the sensitivity of the product's function to the variation in the product parameters.

- Reduce the rate of change of the product parameters.
- Include redundancy.

The most cost-effective approach for reliability improvement is to find appropriate continuous quality characteristics and reduce their sensitivity to all noise factors. Phadke then provides simple examples of a robust design approach. However, in actual application more than one strategy may be necessary.

Uncontrollable user environment factors (such as noise and other stresses) can be planned and organized in an outer array to assess the difference between strength (control factors) and stress. This approach is particularly useful in reliability engineering and robust design, where it is crucial to understand how products perform under varying and often unpredictable real-world conditions.

2.3.5 Understanding the Outer Array

The outer array is a concept borrowed from robust design methodologies, particularly those developed by Genichi Taguchi. It is used to systematically vary uncontrollable factors (also known as noise factors) to understand their impact on the performance of the product or system. Here's how it works:

1) **Identifying Uncontrollable Factors**
   - **Noise Factors:** These are variables that can affect the performance of a product but are not controlled during its use. Examples include environmental conditions like temperature, humidity, and vibration, as well as user behavior such as handling or misuse.
   - **Stress Factors:** These include external pressures like mechanical loads, electrical interference, and other forms of stress that the product might encounter in its operational environment.
2) **Organizing Factors in an Outer Array**
   - **Planning:** Identify all potential uncontrollable factors and their ranges. This involves understanding the real-world environment where the product will be used and the types of stresses it will likely encounter.
   - **Structuring the Array:** Arrange these factors in a systematic way, often using a matrix (outer array) that allows for the controlled variation of each factor. This array is used to test the product under different combinations of these uncontrollable factors.

2.3.6 Assessing Strength vs. Stress

Once the uncontrollable factors are organized in the outer array, the next step is to assess the product's strength (its control factors) against these stresses. Here's how this can be done:

1) Defining Control Factors
   - **Control Factors:** These are variables that can be controlled during the design and manufacturing process. They include material properties, design dimensions, and manufacturing processes that influence the product's strength and performance.
   - **Optimization:** The goal is to identify the optimal levels of these control factors that make the product robust against the variations in uncontrollable factors.

2) Experimental Design and Testing
   - **Conducting Experiments:** Using the outer array, perform a series of experiments where the product is tested under different combinations of noise factors. Measure the product's performance in each scenario.
   - **Data Collection:** Collect data on how the product performs under these varying conditions. This includes measurements of key performance metrics, failure rates, and any degradation in performance.
3) Analysis and Interpretation
   - **Statistical Analysis:** Analyze the data to identify which uncontrollable factors have the most significant impact on performance. Techniques like analysis of variance (ANOVA) can be used to quantify these effects.
   - **Robust Design:** Use the insights gained from the analysis to adjust the control factors, making the product more robust. This might involve changing materials, design tweaks, or altering manufacturing processes to improve resilience against identified stresses.

Benefits and Applications

- **Enhanced Reliability:** By understanding and planning for uncontrollable factors, products can be designed to withstand a wider range of real-world conditions, enhancing their reliability.
- **Cost Savings:** Identifying potential weaknesses early in the design process can reduce the need for costly redesigns and warranty claims.
- **Customer Satisfaction:** Products that perform consistently under various conditions are more likely to meet or exceed customer expectations, leading to higher satisfaction and brand loyalty.

In summary, planning and organizing uncontrollable user environment factors in an outer array to assess the difference between strength (control factors) and stress is a powerful methodology in robust design. It allows for a systematic approach to identify and mitigate the impact of noise factors, leading to products that are more reliable, cost-effective, and capable of meeting the demands of real-world usage.

2.3.7 P-Diagram

In product development, it is beneficial to view the product as a system and utilize the P-diagram (Figure 2.8) to understand the product and identify noise factors. For any product, it is possible to pinpoint DPs, such as material choice, component types, or assembly methods. These parameters have levels that can be set by the system designer. The "P" in P-diagram stands for parameter, process, or product, depending on the context. The core principle of parameter design is to minimize the impact of noise factors on system performance, or responses, by selecting appropriate levels for DPs. This approach is particularly effective when there is a nonlinear relationship between the response and the DPs.

In next section, we will introduce robustness thinking about "how product or process can be designed to insensitive to variation and uncertainty" to explore the synergy between robust engineering and DFR.

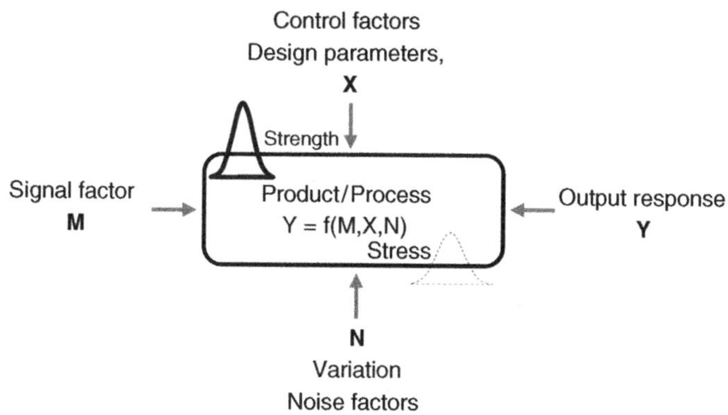

Figure 2.8 P-Diagram.

2.4 Why Robust Design?

Will our customers always use our product under optimal conditions? Will our product always be manufactured under optimal conditions? The answer to both questions is no. The reason is simply that variations exist in products, processes, and everywhere. Customers will not be satisfied if a product or process is sensitive to the effects of variations, as seen in the example of the popcorn-making experience above.

Throughout the years, adopting best practices from the past, numerous leading companies have made substantial investments in the Six Sigma methodology with the primary objective of minimizing waste in manufacturing and operations. These initiatives have significantly influenced the cost structure, consequently impacting the overall financial performance of these companies. As many of them have already realized the maximum potential of the traditional Six Sigma approach, the question arises: What will be the impetus for the next wave of productivity improvement?

Not surprisingly, the ongoing benchmarking survey across numerous leading companies reveals that "design directly influences more than 70% of the product's lifecycle cost. Companies with high product development effectiveness boast earnings three times higher than the average, and those with high product development effectiveness experience revenue growth twice that of the average."

The DFSS approach centers on:

- Enhancing engineering productivity to facilitate the rapid and cost-effective development of new products.
- Implementing value-based management principles.

The robust design method plays a pivotal role in enhancing engineering productivity.

Across a spectrum of industries including automobiles, autonomous driving technologies, electronics, software, and others, companies worldwide have realized significant cost savings, amounting to hundreds of millions of dollars, through the implementation of this method.

2.5 The Importance and Principle of Flow in Robustness Thinking

Robustness thinking, originally developed from Dr. G. Taguchi's principles and enhanced with TRIZ and axiomatic design, is a highly efficient approach to achieving robust design. This methodology has been widely adopted in autonomous solutions, digital technology development, digital transformation, and quality by design. Crucially, robustness thinking should be integrated into the early design phase to ensure the highest level of reliability, as reliability is a key dimension of quality.

Implementing robustness thinking from the start dramatically enhances an organization's competitive edge. Early integration allows for continuous improvement in quality, reliability, and safety, boosting engineering confidence while minimizing costs and eliminating wasteful practices. This proactive approach not only contributes to the bottom line but also ensures that products meet or exceed customer expectations from the outset.

As an example, in 2018, Uber's fatal self-driving car (Figure 2.9) as shown in crash occurred, prompting a significant discussion about the safety of autonomous vehicles. An Uber spokeswoman stated, "We believe that technology has the power to make transportation safer than ever before and recognize our responsibility to contribute to safety in our communities. As we develop self-driving technology, safety is our primary concern every step of the way."

When asked if a Waymo car would have reacted differently than the self-driving Uber involved in the Tempe incident, the CEO of another tech company expressed confidence that its cars would have performed differently under similar circumstances. Waymo's CEO, John Krafcik, emphasized the robustness of their technology. He acknowledged the ongoing investigations and confidently stated, "In situations involving pedestrians, such as a pedestrian with a bicycle, our technology is robust and capable of handling such scenarios." His assessment was based on their understanding of the accident and the robustness built into Waymo's systems.

This incident highlights that the key to safety in autonomous vehicles is not just the integration of multiple new technologies or the fusion of sensors and imaging technologies like radar, lidar, and cameras. The true determinant of safety is the robustness of the technology and the engineered systems based on that technology. Robustness ensures that the system can reliably handle various scenarios and conditions, providing a higher level of safety and performance.

Figure 2.9 Uber's Self-Driving Car, May 29, 2018. *Source:* Adapted from *The Economist*.

By verifying superior robustness in the early design phase, engineers can confidently address potential issues from a mistake-prevention perspective. This foresight enables them to steadily enhance customer satisfaction, knowing that robust design will maintain its performance and reliability over time. Early emphasis on robust design is essential for creating products that are resilient, cost-effective, and capable of delivering consistent quality.

Robustness thinking in design enables the assessment and minimization of inherent design flaws, thereby enhancing the initial quality of the design. By proactively identifying and addressing potential defects early in the design phase, robustness thinking ensures that products are more reliable and consistent from the outset. This approach not only increases the inherent quality of the design but also contributes to greater customer satisfaction and reduced costs associated with correcting defects later in the production process.

Robustness thinking is not confined to engineering, production plants, or manufacturing companies. This approach can benefit any business, including those in the supply chain, that aims to produce high-quality products and deliver exceptional services at the lowest cost while minimizing defects and waste. Whether it's a restaurant, an automotive manufacturer, an electronics company, or a business focused on digital technologies and digital transformation, all can adhere to the same five principles of robust design derived from Taguchi's method. These principles are:

1) **Identify and Select Proper System Output Response from the Energy Flow Perspective:** Understand the system's energy flows to identify the most appropriate output responses, ensuring the system's overall effectiveness and efficiency.
2) **Measure Function Using S/N Ratio:** Utilize the signal-to-noise (S/N) ratio to measure and improve the robustness of the system's functions, enhancing quality and reliability by reducing variability.
3) **Take Advantage of Interactions between Control and Noise Factors:** Leverage the interactions between controllable and uncontrollable factors to optimize system performance under varying conditions, thus improving robustness.
4) **Use Orthogonal Arrays:** Employ orthogonal arrays in experimental design to systematically and efficiently study the effects of multiple factors on performance, leading to better decision-making and design improvements.
5) **Two-Step Optimization:** First, optimize for robustness by focusing on the design's insensitivity to noise factors, and then optimize for performance to achieve the best possible results under given constraints.

In addition to these principles, robustness-thinking-based activities foster and facilitate innovation by decoupling coupled designs and manage interactions to improve consistent performance. By applying these principles and activities, businesses can:

- **Innovate Through Decoupling:** Identify and decouple interactions within the design to simplify complexity and enhance flexibility. This process reduces the impact of any single factor's variability on the overall system, fostering innovation by allowing independent optimization of different components.
- **Facilitate Consistent Performance:** Use robustness thinking to design systems that maintain consistent performance despite variations in external conditions. This reliability encourages a culture of innovation, as teams can confidently experiment with new ideas without compromising the system's stability.

- **Encourage Proactive Problem-Solving:** Implement early detection and mitigation strategies for potential design flaws, leading to more resilient and innovative solutions. This proactive approach ensures that the design process remains agile and adaptable to new challenges and opportunities.
- **Promote a Culture of Continuous Improvement:** Embrace a mindset of ongoing improvement and learning, where every iteration of the design process builds on previous successes and lessons learned. This culture drives innovation by constantly pushing the boundaries of what is possible.

The focus of this approach is on flow—understanding the three flows of transformation, identifying barriers to flow in design, and implementing strategies to improve flow. Robustness thinking in design enables the assessment and minimization of inherent design flaws, thereby enhancing the initial quality of the design. By proactively identifying and addressing potential defects early in the design phase, robustness thinking ensures that products are more reliable and consistent from the outset. This approach not only increases the inherent quality of the design but also contributes to greater customer satisfaction and reduced costs associated with correcting defects later in the production process.

By following these principles and engaging in robustness thinking-based activities, businesses across various industries can significantly improve their competitive edge, ensure consistent quality, foster innovation, and enhance customer loyalty, all while maintaining cost-effectiveness and operational efficiency.

2.5.1 Defining Flow

Flow refers to the process by which inputs—such as materials, information, and energy—are transformed within a system to produce the intended outputs. When a system exhibits "good" flow, it operates smoothly and predictably, consistently delivering the desired outcomes. Conversely, "bad" flow is characterized by disruptions or inefficiencies, leading to both hard and soft failure modes. Every breakdown in flow increases the likelihood of accumulating waste, which can hinder overall system performance.

Achieving optimal flow is crucial for maximizing efficiency and effectiveness. The goal is to maintain a steady and consistent flow, ensuring that the system consistently delivers high-quality, reliable functionality. This, in turn, generates greater value for customers by providing products and services that meet or exceed their expectations.

Key Aspects of Flow:

- **Smooth Transformation:** Inputs should move through the system with minimal resistance or interruption, transforming efficiently into the desired outputs.
- **Predictability:** A well-functioning system with good flow operates predictably, with processes that can be consistently repeated to achieve the same high-quality results.
- **Waste Minimization:** By maintaining a steady flow, systems reduce the accumulation of waste, whether in the form of excess materials, time, or energy.
- **Reliability:** Consistent flow leads to more reliable system performance, as processes are optimized to avoid breakdowns and failures.
- **Customer Value:** Systems with optimized flow generate greater value for customers by delivering products and services that are both high in quality and reliable.

By focusing on enhancing flow, businesses can improve their operational efficiency, reduce waste, and deliver better value to their customers. This approach not only boosts overall system performance but also contributes to long-term success and sustainability.

2.5.2 Transformation Systems, Flow, and Proactive Failure Creation

Engineering systems are designed to deliver their basic functions by utilizing one or more of three types of transformations: material, energy, and signal transformations. According to the discussions by Pahl & Beitz and Hubka, an energy-related system output response is crucial because it aligns with the fundamental purpose of an engineering system.

Engineering systems exist to transform:

- **Energy**
- **Material**
- **Information**

from input to output.

Transformation Creates Loads

Any transformation involves a change in state. That change requires physical effort, which manifests as **loads**. If a function exists, loads exist.

Examples include:

- Electrical-to-mechanical conversion creating torque and heat
- Information (control logic) modulating energy and creating dynamic loads
- Material movement creating inertia, pressure, and contact forces

Failures cannot occur unless loads exist. Therefore, proactive reliability begins by **explicitly identifying how system functions create loads**, rather than starting with observed failures.

The Three Flows of a System:

1) **The Flow of Raw Material:** This involves the movement and transformation of physical substances within the system.
2) **The Flow of Information:** This includes the transfer and processing of data and signals to control and optimize system performance.
3) **The Flow of Energy:** This refers to the distribution and conversion of energy forms necessary for the system to function.

2.5.2.1 Load–Stress–Strength Thinking as a Proactive Reliability Framework

Load Load represents the external demand placed on a system by:

- Customer usage
- Environment
- Mission profile
- Interactions with other systems

Loads are requirements, not defects.

Stress Stress is the internal response of the design to applied load. Stress depends on:

- Geometry
- Materials
- Interfaces
- Architecture

Load alone does not cause failure.

Strength Strength is the designed capability of the system to withstand stress without failure. Strength is created through:

- Material selection
- Geometry and load paths
- Architecture and redundancy
- Control limits and protections

Strength is designed in, not discovered through testing. Failure occurs only when **stress exceeds strength**, and proactive reliability focuses on increasing the separation between them under variation.

2.5.2.2 Margin, Limits, and Failure Distance

Traditional margin-based design compares nominal stress to nominal strength. However, failures occur due to **variation**, not averages.

Robustness thinking introduces **failure distance**, defined as the separation between stress and strength **distributions**, not single values.

Proactive reliability therefore evaluates:

- Stress mean and variability
- Strength mean and variability
- Overlap probability

This distribution-based view enables early, analytical failure avoidance.

2.5.2.3 Noise Factors and Robust Design for Proactive Reliability

Noise factors are unavoidable and include:

1) Manufacturing piece-to-piece variation
2) Aging and degradation
3) Customer usage variability
4) External environment
5) Interfaces and interactions

Robust design optimizes **control factors** so that performance is insensitive to these noise factors. This approach:

- Maximizes effective strength
- Reduces stress amplification
- Minimizes variability of functional response
- Prevents harmful side effects of noise

Robust design is therefore the **primary mechanism for proactive reliability creation**.

2.5.2.4 Architecture Robustness and Failure Propagation

Architecture decisions dominate reliability outcomes. Robust architectures:

- Eliminate single-point failures
- Limit failure propagation
- Enable graceful degradation
- Preserve safe functionality under fault conditions

Proactive reliability is achieved when architectural robustness prevents local failures from becoming system-level failures.

2.5.2.5 Reliability Creation During Concept and Design

Reliability is established primarily during early design phases. Robustness thinking enables proactive reliability by:

- Eliminating high-risk concepts early
- Selecting inherently robust architectures
- Designing intentional load paths
- Reducing sensitivity to variation before tolerance tightening

This approach minimizes late-stage rework and lifecycle cost.

2.5.2.6 Summary: Robustness Thinking as Proactive Reliability

Robustness thinking provides the foundation for proactive reliability by:

- Managing variation and uncertainty
- Designing stress–strength separation
- Creating architecture resilience
- Enabling failure avoidance by design

Reliability is not tested into a product,
Reliability is created through robust design.

2.5.3 Importance of Flow in System Design and Optimization

For designing, evaluating, and optimizing an engineered system or product, it is essential to select an appropriate system output response that evaluates and comprehends the system's functional behavior. This response should be related to the system's basic function, which is the transfer of material, energy, and information from the system's input to its output. Essentially, the basic function of a product or process technology is its ability to transform input into output in terms of material, information, and energy.

2.5.4 Integrating Robustness Thinking and Robust Design Principles

Companies that successfully integrate robustness thinking and the principles of robust design understand that harmonious transformation of material, energy, and information flows leads to optimal product function without symptoms or side effects, such as soft or hard failures. Maintaining smooth product performance or production processes ensures the creation of both effectiveness and efficiency.

Benefits of Harmonious Flow

- **Efficiency:** Harmonized flow reduces waste and minimizes deviations, ensuring processes run smoothly.
- **Reliability:** Consistent and predictable system performance enhances overall reliability.
- **Customer Value:** High-quality, reliable outputs increase customer satisfaction and loyalty.
- **Profitability:** Enhanced efficiencies improve overall business performance, leading to increased profitability.

Companies implementing lean manufacturing and robustness thinking can achieve significant competitive advantages by optimizing the three flows of transformation. These practices not only improve operational efficiency but also contribute to greater profitability, setting these companies apart from their competitors.

2.5.5 Barriers to Flow Due to Lack of Robustness Thinking

The absence of robustness thinking in the design and operation of systems can lead to significant barriers to flow. These barriers can manifest in various forms, disrupting the smooth transformation of materials, energy, and information within the system. Here are several key barriers that can arise from not incorporating robustness thinking:

1) **Increased Variability:** Without robustness thinking, systems are more susceptible to variations in input materials, environmental conditions, and operational parameters. This increased variability can lead to inconsistent outputs, frequent adjustments, and a higher likelihood of defects, all of which disrupt the flow.
2) **Unanticipated Failures and Safety:** A lack of robustness can result in systems that are not resilient to unexpected changes or stresses. This vulnerability can cause unanticipated failures, leading to downtime, costly repairs, and interruptions in the flow of production or services.
3) **Suboptimal Performance:** Systems not designed with robustness in mind may struggle to maintain optimal performance under varying conditions. This can lead to inefficiencies, increased waste, and higher operating costs, all of which impede the smooth flow of processes.
4) **Increased Waste:** Systems that are not robust tend to generate more waste due to higher defect rates, rework, and inefficient use of resources. This waste accumulation can clog the flow, reducing overall system efficiency and effectiveness.
5) **Complexity in Operations:** Lack of robustness can result in more complex operational requirements, as additional controls and adjustments are needed to compensate for system weaknesses. This complexity can slow down processes and create bottlenecks in the flow.
6) **Reduced Customer Satisfaction:** Systems that fail to deliver consistent, high-quality outputs due to a lack of robustness can lead to dissatisfied customers. Poor reliability and performance can erode customer trust and loyalty, further impacting the smooth flow of business operations.

7) **Increased Costs:** Addressing the consequences of nonrobust systems, such as frequent maintenance, repairs, and quality control measures, can significantly increase operational costs. These added expenses can strain resources and disrupt the financial flow of the organization.

2.5.6 Overcoming Barriers to Flow with Robustness Thinking

To overcome these barriers and enhance the flow within systems, organizations should integrate robustness thinking into their design and operational processes. Here are several strategies to achieve this:

- **Implement Robust Design Principles:** Focus on designing systems that are insensitive to variations and capable of maintaining performance under diverse conditions. Use principles such as the S/N ratio, orthogonal arrays, and two-step optimization to achieve robust designs.
- **Early Detection and Mitigation:** Identify potential sources of variability and failure early in the design phase. Implement measures to mitigate these risks and enhance system resilience from the outset.
- **Continuous Improvement:** Foster a culture of continuous improvement where feedback from operations is used to refine and enhance system robustness. Regularly assess and update designs to address emerging challenges and opportunities.
- **Lean Manufacturing Practices:** Integrate lean manufacturing principles to streamline processes, reduce waste, and improve overall flow. Focus on creating value by eliminating nonessential activities and optimizing resource use.
- **Training and Education:** Invest in training and educating employees on robustness thinking and its importance in achieving smooth flow. Equip teams with the knowledge and tools to identify and address barriers to flow effectively.

By incorporating robustness thinking into their operations, organizations can overcome barriers to flow, enhance system performance, and deliver greater value to customers. This approach not only improves operational efficiency but also contributes to long-term sustainability and competitive advantage.

2.5.7 Examples of Barriers to Flow

Two examples below illustrate the barriers to flow due to the lack of robustness thinking:

1) **Manufacturing Defects and Rework:** In a manufacturing environment, the lack of robustness in the design of a production process can lead to high defect rates. For example, a car assembly line might experience frequent issues with part alignment due to variations in material quality or environmental conditions. This results in a high number of defective units, which require rework or scrapping. The need for rework disrupts the smooth flow of production, increases waste, and adds to operational costs. By implementing robustness thinking and designing processes that are less sensitive to these variations, the manufacturer can reduce defects, ensure a smoother flow, and improve overall efficiency.

2) **Software System Failures:** In the development of a software system, a lack of robustness can lead to frequent system crashes or bugs when users interact with the software under different conditions. For instance, an e-commerce platform may experience downtime during peak usage periods due to inadequate handling of high traffic loads. This disruption in the flow of information can result in lost sales, frustrated customers, and damage to the company's reputation. By applying robustness thinking in the software design phase, developers can create systems that are more resilient to varying loads and user interactions, ensuring a smoother flow of information and a better user experience.

2.5.8 Addressing Barriers with Robustness Thinking

To overcome these barriers and enhance the flow within systems, organizations should integrate robustness thinking into their design and operational processes. Here are several strategies to achieve this:

- **Implement Robust Design Principles:** Focus on designing systems that are resilient to variations and capable of maintaining performance under diverse conditions. Use principles such as the S/N ratio, orthogonal arrays, and two-step optimization to achieve robust designs.
- **Early Detection and Mitigation:** Identify potential sources of variability and failure early in the design phase. Implement measures to mitigate these risks and enhance system resilience from the outset.
- **Continuous Improvement:** Foster a culture of continuous improvement where feedback from operations is used to refine and enhance system robustness. Regularly assess and update designs to address emerging challenges and opportunities.
- **Lean Manufacturing Practices:** Integrate lean manufacturing principles to streamline processes, reduce waste, and improve overall flow. Focus on creating value by eliminating nonessential activities and optimizing resource use.
- **Training and Education:** Invest in training and educating employees on robustness thinking and its importance in achieving smooth flow. Equip teams with the knowledge and tools to identify and address barriers to flow effectively.

By incorporating robustness thinking into their operations, organizations can overcome barriers to flow, enhance system performance, and deliver greater value to customers. This approach not only improves operational efficiency but also contributes to safety, long-term sustainability and competitive advantage.

"Robust engineering commences with 'thinking in terms of function,' specifically, the conversion of input into a desired output. It acknowledges that systems transform energy from one form to another to fulfill an 'intended' function."

These energy transformations adhere to the laws of physics, although there is always some loss of energy and variability caused by uncontrollable variables such as operating conditions and aging (Figure 2.10).

The rationale for utilizing an energy-related system output response, as discussed by Pahl and Beitz [5] and Hubka, stems from the inherent nature of engineering systems, which are always designed to fulfill their basic functions. To achieve their basic functions, at

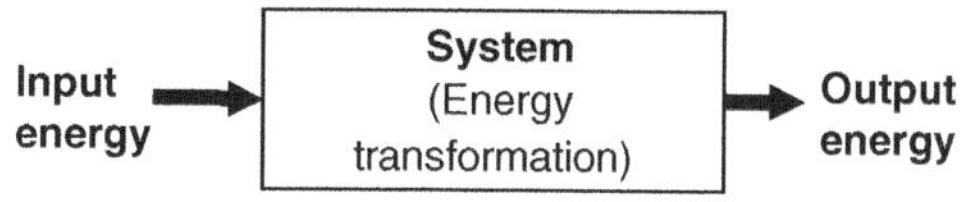

Figure 2.10 Energy Transformation.

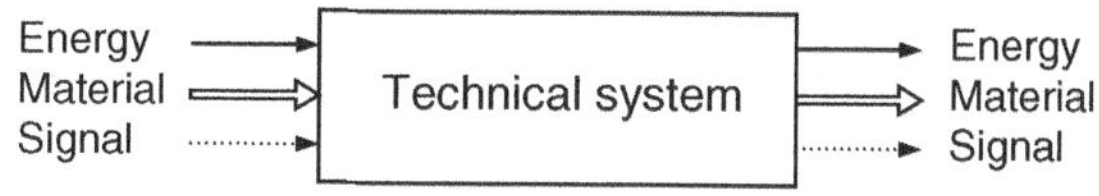

Figure 2.11 Technical System: The Conversion of Energy, Material, and Signals.

least one of the three types of transformation—material, energy, and signal transformation (Figure 2.11)—must be employed.

Details of energy, material, and signal include:

- Energy: Mechanical, thermal, electrical, chemical; also force, current, heat.
- Material: Liquid, gas; also, raw material, end product, component.
- Signal: Information, data, display, magnitude.

For example, in a machining process, viewed as an engineering system, the ideal relationship between output and input should entail the output dimensions precisely matching the intended dimension. This type of transformation system is categorized as material transformation. Given the significance of energy transformation, there are notable similarities in utilizing these three types of transformations to identify the appropriate output characteristic.

2.6 Robustness Development Strategy

In the realm of robustness development, there's a guiding principle: "Instead of measuring reliability directly, focus on evaluating the robustness of design functionalities." What does this entail?

Rather than simply tallying the number of failures to understand system malfunctions or enhance them, the emphasis lies on gauging the variability in the functional responses and comparing it against the requirements of the function. Essentially, it suggests that merely counting failures and their frequency does not offer significant insight into their underlying causes or how to rectify them. By solely concentrating on failure count, one overlooks parameters crucial for predicting or preventing failures. Although failure rate holds importance for customers and service, it primarily serves as a trailing indicator for engineering.

Through the application of solid scientific principles and engineering practices, robustness is initially achieved without resorting to stricter tolerances, higher-quality materials, or other strategies that inflate manufacturing or service expenses. By subjecting systems to rigorous laboratory conditions, issues can be identified swiftly, and superior solutions can be devised. Implementing this approach during the early development stages helps evade costly modifications to designs and tooling later on, thus aligning with the objective of "lean" product development. When robustness is assessed in terms of variability under deliberately stressful conditions, it serves as an early indicator of potential deterioration under real-world stressors, making it more sensitive to driving design enhancements.

Consequently, prioritizing variation is deemed more crucial for reliability development than fixating on failure occurrences and their rates. The subsequent discussion outlines the methodology for measuring reductions in variability within the context of assessing improvements in robustness under controlled laboratory conditions. While customers may perceive it as aiming for a more reliable or durable product, fundamentally, enhancing the robustness of a product's design is the pathway to achieving greater reliability and availability.

Variation reduction is universally recognized as a key to reliability and productivity improvement. There are many approaches to reducing variability, each one having its place in the product development cycle.

By addressing variation reduction at a particular stage in a product's lifecycle, one can prevent failures in the downstream stages. The Six Sigma approach has made tremendous gains in cost reduction by finding problems. The robustness strategy is to prevent problems through optimizing product designs and manufacturing process designs.

The manufacturer of a wire-terminal assembly used in automotive connectors faced the problem of long-term reliability due to manufacturing variability. Holes in the assembly and low pull strength caused poor long-term reliability. So, how to minimize field problems and associated cost? There are many approaches:

1) Compensate the customers for their losses.
2) Screen out assembled wire-terminal having holes and low pull strength at the end of the production line.
3) Institute tighter tolerances through process control on the manufacturing line.
4) Change the nominal values of critical wire-terminal parameters so that the assembled wire-terminal function becomes insensitive to the causes, such as manufacturing variation and user conditions.

Approach 4 is the robustness strategy. As one moves from approach 1 to 4, one progressively moves upstream in the product delivery cycle and also becomes more efficient in cost control. Hence it is preferable to address the problem as upstream as possible. The robustness strategy provides the crucial methodology for systematically arriving at solutions that make designs less sensitive to various causes of variation. It can be used for optimizing product design as well as for manufacturing process design.

After a concept design is generated, there are some different approaches to achieve product design robustness. These approaches include the following:

- Design of Experiments (DOE—bring the nominal to target).
- Failure Mode and Effects Analysis (FMEA—detect and prevent failure modes).
- Tolerance Design (spending money to improve manufacturing process capability).

In addition to these approaches, robust parameter design focuses on first reducing performance variation and then achieving target performance. It starts with asking: What is the intended function of the system? What is the associated energy transformation? How is it best measured? Robust parameter design then searches for a design that minimizes variability of the intended function, regardless of the manufacturing or end-user environment. Robust parameter design is one element of robustness strategy.

In general, engineering decisions involved in product/system development can be classified into two categories:

- Error-free implementation of the past collective knowledge and experience.
- Generation of new design information, often for improving product quality/reliability, performance, and cost.

While CAD/CAE tools are effective for implementing past knowledge, robust design method greatly improves productivity in generation of new knowledge by acting as an amplifier of engineering skills. With robust design, a company can rapidly achieve the full technological potential of their design ideas and achieve higher profits.

2.7 Three Phases of Robust Design

There are three robust design phases in the design of a product or manufacturing process. They are as follows:

- Concept design.
- Parameter design.
- Tolerance design.

These three phases are designed to build robustness into a product or process in contrast to an iterative design/build/test/fix approach. While parameter design can also be an iterative approach, it is a more systematic means of reducing performance variability.

Taguchi's approach to design emphasizes continuous improvement and encompasses different aspects of the design process grouped into three main stages:

- **System design**. This broadly corresponds to conceptual design in the generalized model of the design process. System design is the conceptual design stage in which scientific and engineering expertise is applied to develop new and original technologies. Robust design using Taguchi method does not focus on the system design stage.
- **Parameter design.** Parameter design is the stage at which a selected concept is optimized. Many variables can affect a system function. The variables need to be characterized from an engineering viewpoint. The goals of parameter design are to (1) find that combination of control factors settings that allow the system to achieve its ideal function and (2) remain insensitive to those variables that we cannot control. Parameter design provides opportunities to reduce the product and manufacturing costs.
- **Tolerance design.** Although generally considered to be part of the detail design stage, Taguchi views this as a distinct stage to be used when sufficiently small variability cannot be achieved within a parameter design. Initially, tolerances are usually taken to be fairly wide because tight tolerances often incur high supplier or manufacturing costs. Tolerance design can identify those tolerance that, when tightened, produce the most substantial improvement in performance.

The result of using parameter design followed by tolerance design is successful products at low cost.

As illustrated in Figure 2.12, these three phases support several elements in the upstream identify, design, optimize, verify (IDOV) DFSS process. Concept design is part of the Design phase, and parameter and tolerance design are part of the Optimize phase.

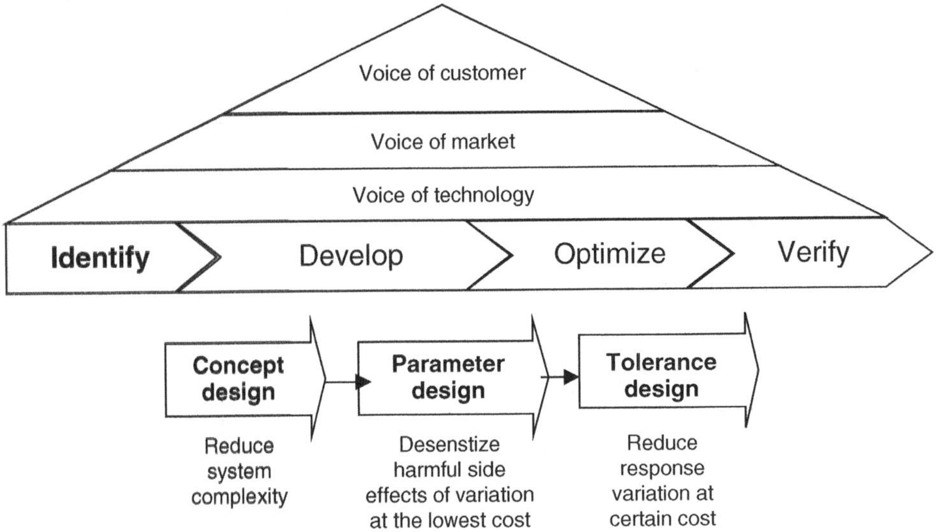

Figure 2.12 Three Phases of Upstream.

System design is the conceptual design stage in which scientific and engineering expertise is applied to develop new and original technologies. Conceptual design is the first phase of product development. It is here that the concepts to be included in a product are developed. Product development is different from technology development. Technology development is the process of conducting research and development around fundamental phenomenological building blocks. This is where the laws of physics are uncovered and/or explored for potential commercial applications. It is where what can happen is discovered, documented, and certified as a viable technology that can be made to produce repeatable results. Once a technology is certified for its stability of behavior and general susceptibility to various sources of variation, it is then ready to be used in the first phase of product development-conceptual design.

One of conceptual design tasks is used to compare various competing technologies that have been proved to be mature enough to safely enter commercialization. Conceptual design is a form of benchmarking, which usually compares one or more existing products against other competing products from the marketplace. A key thing to remember about concept design is that it is intended to reject any concepts based on immature technologies or design structure that are trying to creep into the commercialization process.

The key point to remember about Figure 2.5 is that parameter design is the product or process design phase in which the engineer, through structured experimentation and Two-step optimization, identifies the best nominal levels for DPs to reduce response variation in the presence of noise followed by adjusting the response to target.

Parameter design is a process that will be employed by the multifunctional engineering team to optimize the superior concept against the noises inherent in the fabrication and the use of the design. The value in doing concept design first in that one gets to pick the noises that are going to have to be accommodated during the rest of the product development process. It is much easier to perform parameter design on a superior concept than it

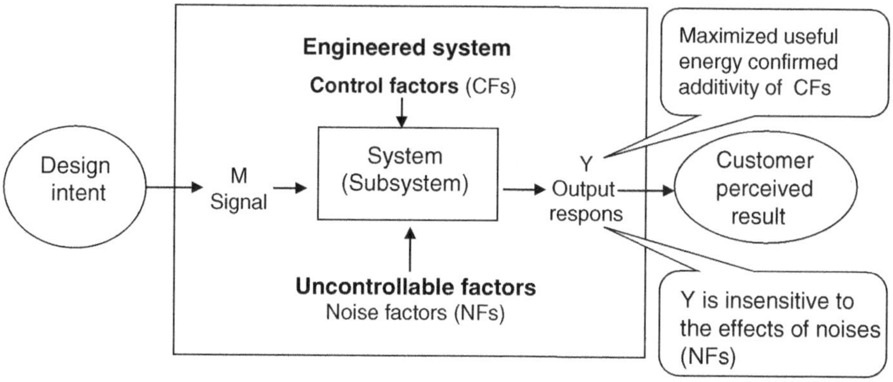

Figure 2.13 The Engineered System and Parameter Design.

is to try to improve a design that is laced with sensitivities to noises and functional structure (functional coupling and interactions) that may not be readily apparent to the current engineering team. The design with sensitivities to noises may have only minimal optimization capability—it all depends on the physics underlying the design and the strength of the interaction between the engineering parameters and the noise factors. This is the reason why the concept design stage is so important to develop an inherent robust design concept.

Parameter design is the stage at which a selected concept is optimized. Many variables can affect a system function. The variables need to be characterized from an engineering viewpoint. Some of the variables will be controllable (control factors). The engineer should set the nominal values of these variables based on known or anticipated nonlinear relationships between levels of control factors and selected performance measures. Others will be uncontrollable or too expensive to control (noise factors). See P-diagram for more explanation for noise factors. These are the downstream variables that cause functional variation (see Figure 2.13).

The goals of parameter design do not include finding and removing these causes of variation or upgrading material and components to improve quality. The goals are to:

- Find that combination of control factors settings that allow the system to achieve its ideal function.
- Remain insensitive to those variables that we cannot control.

This allows engineers to develop designs with high performance, stability, and reliability. The stability of performance with the side effects of noise factors is termed robustness, robust design, or robust quality.

2.8 Understanding and Mitigating Mistakes in Design and Manufacturing

By "mistake," we refer to the wide range of design decisions and manufacturing operations that can be significantly erroneous. Examples of such mistakes include:

1) **Installation Errors:** For instance, installing a switch backward, which can lead to malfunction or safety hazards.

2) **Misinterpretation of Specifications:** An example is interpreting a software command as being in inches when it is actually in centimeters, leading to incorrect operations or measurements.

2.8.1 Improving Reliability by Reducing Mistakes

Reliability can be significantly enhanced by minimizing the occurrence of these mistakes. This can be achieved through a combination of knowledge-based engineering and a structured problem-solving process. Let us delve deeper into these strategies:

Knowledge-based Engineering

1) **Standardization:** Implementing standardized procedures and guidelines can help ensure consistency and reduce the likelihood of errors. For example, using standardized units of measurement across all software and hardware components can prevent misinterpretation.
2) **Training and Education:** Providing comprehensive training programs for engineers and manufacturing personnel ensures that they are well versed in the latest technologies, methodologies, and best practices.
3) **Documentation and Checklists:** Maintaining detailed documentation and checklists for design and manufacturing processes can serve as a reference and verification tool, reducing the chances of oversight.
4) **Design Reviews and Audits:** Conducting regular design reviews and audits can help identify potential errors early in the development process, allowing for corrective actions before production.

Problem-solving Process

1) **Root Cause Analysis:** When a mistake occurs, performing a thorough root cause analysis helps identify the underlying factors contributing to the error. Techniques such as the "5 Whys" or fishbone diagram can be useful.
2) **Corrective and Preventive Actions (CAPA):** Implementing CAPA ensures that once a mistake is identified, steps are taken to correct it and prevent its recurrence. This might involve process changes, additional training, or design modifications.
3) **Continuous Improvement:** Adopting a culture of continuous improvement, such as through Lean or Six Sigma methodologies, encourages ongoing evaluation and enhancement of processes to reduce errors.
4) **Simulation and Testing:** Utilizing advanced simulation tools and rigorous testing protocols can help identify potential mistakes in the design phase, allowing for adjustments before manufacturing begins.

Technological Aids

1) **Automation:** Leveraging automation in both design and manufacturing processes can reduce human error. For example, automated assembly lines can ensure components are installed correctly.

2) **Advanced Software Tools:** Using sophisticated design and analysis software can help detect inconsistencies and potential errors early in the design process.
3) **Real-Time Monitoring:** Implementing real-time monitoring systems in manufacturing can quickly identify deviations from the norm, allowing for immediate corrective actions.

Collaborative Efforts

1) **Cross-functional Teams:** Encouraging collaboration between design, engineering, and manufacturing teams can lead to a better understanding of potential pitfalls and collective problem-solving.
2) **Supplier Partnerships:** Working closely with suppliers to ensure they understand the specifications and quality requirements can reduce errors in the supply chain.

By integrating these strategies, organizations can create a robust framework that not only reduces the incidence of mistakes but also fosters a culture of quality and reliability. This holistic approach ensures that products are designed and manufactured to meet high standards, ultimately leading to greater customer satisfaction and reduced costs associated with errors and rework.

Failure-mode avoidance is a proactive design strategy that aims to expand the operational boundaries of a system, creating a larger region in which it can function effectively. This approach goes beyond merely meeting specified operating conditions and seeks to create inherently robust systems.

References

1 Hu, M. (2014). Robustness thinking in design for reliability, RAMS. *Annual RELIABILITY and MAINTAINABILITY Symposium*.
2 Box, G.E.P., Bisgaard, S., and Fung, C. (1988). An explanation and criticism of Taguchi's contribution to quality engineering. *Quality and Reliability Engineering International* 4: 123–141.
3 Box, G.E.P. and Jones, S. (1992). Split-plot designs for robust experimentation. *Journal of Applied Statistics* 19: 3–25.
4 Phadke, M.S. (1989). *Quality Engineering Using Robust Design*. Englewood Cliffs, NJ: Prentice-Hall.
5 Pahl, G. and Beitz, W. (1995). *Engineering Design, A Systematic Approach*, 2nde. Spinger.

Further Reading

Bergman, B. (2009). *Robust Design Methodology for Reliability - Exploring the Effects of Variation and Uncertainty* (ed. J. de Maré, S. Lorén, and T. Svensson). Chichester: Wiley.

Chakhunashvili, A., Johansson, P., and Bergman, B. (2004). Variation mode and effect analysis. In: *Proceedings of the Annual Reliability and Maintainability Symposium*. Los Angeles, CA: IEEE.

Clausing, D. (1994). *Total Quality Development. A Step-By-Step Guide to World-Class Concurrent Engineering*. New York: ASME Press.

Davis, T.P. (2006). Science, engineering, and statistics. *Applied Stochastic Models in Business and Industry* 22: 401–430.

Gamez-Montero, P.J. and Bernat-Maso, E. (2022). Taguchi techniques as an effective simulation-based strategy in the design of numerical simulations to assess Contact Stress in Gerotor Pumps. *Energies* 15: 7138.

3

Robust Design Principles, Tactics, and Primary Tools

CHAPTER MENU

3.1 Introduction

A robustness strategy is a comprehensive plan for achieving robustness, encompassing objectives, goals, and paths. Tactics are the specific means for executing this strategy, utilizing particular tools to achieve the established objectives, goals, and paths. The key question is: Where and how can we achieve design robustness at the lowest possible costs?

Robust design is a methodology focused on enhancing product quality and reliability by minimizing the impact of variability, especially uncontrollable environmental factors ("noise"). This approach ensures consistent product performance under diverse conditions. Implementing robust design principles involves using specific tools and techniques to systematically improve product reliability and performance. Here are the core principles and primary tools of robust design:

- Consistency: Ensuring reliable product performance across various conditions.
- Minimizing Variability: Reducing the impact of uncontrollable factors on product performance.
- Systematic Improvement: Using structured methods to enhance reliability and performance.
- Robust Parameter Design Tactics: Efficient methods for achieving robustness.
- Primary Tools: Specific instruments to effectively meet design goals.

Robust Design Tactics:
Tactics are the means by which the robust design strategy is executed, involving the use of specific tools and techniques to achieve the set objectives, goals, and paths. The primary tactics for robust parameter design are:

1) **Parameter Selection:** Identifying and choosing the right parameters that influence product performance.
2) **Parameter Optimization:** Fine-tuning parameters to achieve the best possible performance with minimal variability.
3) **Experimental Design:** Conducting systematic experiments to understand the effects of various parameters and noise factors.

Primary Tools:
To implement the robust design tactics effectively, several primary tools are used:

1) **Design of Experiments (DOE):** A statistical method to plan, conduct, analyze, and interpret controlled tests to evaluate the factors that control the value of a parameter or a group of parameters.
2) **Parameter Design:** Techniques for improving product quality by designing experiments that can systematically identify and desensitize the impact of noise factors.
3) **Tolerance Design:** Methods to determine acceptable limits of variability in product dimensions and performance attributes at affordable cost.

By leveraging these principles, tactics, and tools, robust design aims to achieve high reliability and quality in products while minimizing costs and the impact of uncontrollable variability.

3.2 Ideal Function: Ideal Transformation System Input and Output Relationship

The ideal function is employed to mathematically define the ideal form of the signal–response relationship, as encapsulated by the design concept aiming to make the higher-level system function perfectly. Establishing this ideal relationship is an engineering or technological task, and it is not the primary objective of statistical analysis or experimental design methods. Nevertheless, statistical analysis and experimental design can be used inductively to assist in the development of the ideal function.

According to Taguchi, it is crucial to avoid using "quality symptoms" such as reliability data, warranty information, scrap, and the percentage of defective items as output characteristics in the late stages of product development and manufacturing environments. Improving a "symptom" may not effectively enhance the robustness of the system's ability to deliver its functions, which is the primary objective of a robust design project. Understanding the system's function, especially its basic function, is key to robust technology development [1]. Defining the ideal state of the basic function, referred to as the ideal function, serves as the centerpiece for identifying the output characteristic [2, 3].

Achieving efficiency in the transformation of energy, wherein an ideal relationship between input energy and output response is attained, represents the strategic thinking supporting the ideal function concept. An engineered system attains its "ideal function" when all applied energy (input) is efficiently transformed into the creation of desired output energy. If such an ideal function exists in reality, there would be no energy losses resulting in symptoms of poor function, such as noise, scrap, rework, and quality concerns.

3.3 Ideal Function and Quality Problems

Quality problems arise when a system's design intent cannot be realized. An illustration of the ideal function in image design and quality issues, as presented by Mori, is used for clarification. Let us consider A1 and A2 combinations of a photosensitive material. Figure 3.1 illustrates the relationship between the hardness of the photosensitive material after exposure and the light energy used for exposure. Light energy levels E1 and E2 are required to achieve sufficient hardness (H) through the curing process facilitated by photoreaction.

Now if we assume that all the energy of emitted light E1 is consumed by the photoreaction of A1, light energy E1 is the useful energy that achieves necessary hardness H.

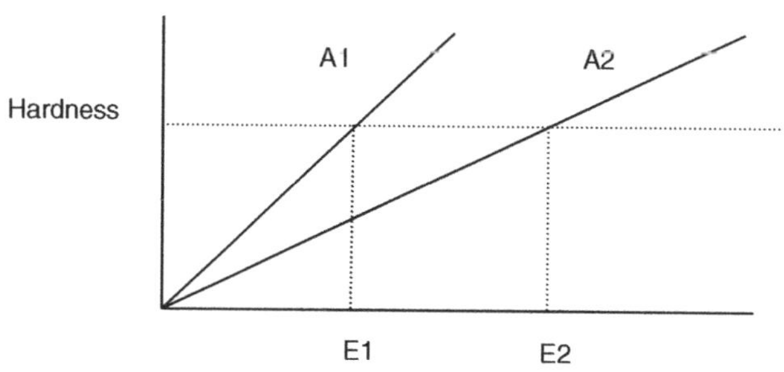

Figure 3.1 Hardness (H) and Light Energy (E).

E2, on the other hand, is the energy emitted to achieve hardness H for A2. However, because E1 is the sufficient energy, E2–E1 is the amount of redundant energy.

$$\begin{aligned} E1 &= \text{useful component} \\ E2 &= E1 + (E2 - E1) \\ &= \text{useful component} + \text{harmful component} \\ &= \text{signal component} + \text{noise component} \\ &= S + N \end{aligned} \tag{3.1}$$

If E2–E1 does not impact any other performance, it is merely a loss of energy, and there is no quality-related problem. However, excessive light energy, in general, becomes a cause of halation, leading to an undesired growth of the image. Furthermore, as light energy is converted into heat energy inside the photosensitive material, interface reactions with the base material occur, causing forging. Therefore, since excessive light energy (E2–E1) serves as an energy source that generates side effects, it becomes a harmful component. Consequently, light energy level E2 is decomposed into a useful component (E1) and a harmful component (E2–E1).

In robust engineering, the former is referred to as the signal component, denoted by S, and the latter as the noise component, denoted by N. The evaluation of photosensitive material combinations is conducted using a figure obtained by dividing the signal component S by the noise component N. This ratio is known as the signal-to-noise (S/N) ratio.

$$\begin{aligned} \text{S/N Ratio} &= \text{Signal Component(S)/Noise Component(N)} \\ &= \text{Energy(Power) that is transformed into intended output/} \\ &\quad \text{Energy(Power) that is transformed into unintended output} \end{aligned} \tag{3.2}$$

If the signal component S is large and the noise component N is small, the S/N ratio becomes large. Conversely, a condition that yields a small signal component S and a large noise component N, resulting in a significant side effect, produces a small S/N ratio. Therefore, the judgment of whether a condition is favorable can be made using the S/N ratio. The S/N ratio becomes large when the condition is close to the ideal function. Further details on the S/N ratio will be explained in Section 3.7 later.

3.4 Identification and Classification of Design Parameters: P-Diagram

The P-diagram is highly useful in every robust design and Design for Six Sigma (DFSS) and robust design for reliability (RDfR) project as it succinctly defines the scope of robust design or RDfR. Initially, we identify the signal (input) and response (output) associated with the design concept. For instance, in designing the weight scale system, as shown in Figure 3.2, for an object, the signal is the object being weighed, and the resulting weight measurement is the response.

Next, consider the parameters/factors that are beyond the control of the designer, known as noise factors. Examples of noise factors include outside temperature and user conditions. On the other hand, parameters that can be specified by the designer are called control factors, with examples such as the sizes of the scale and spring force.

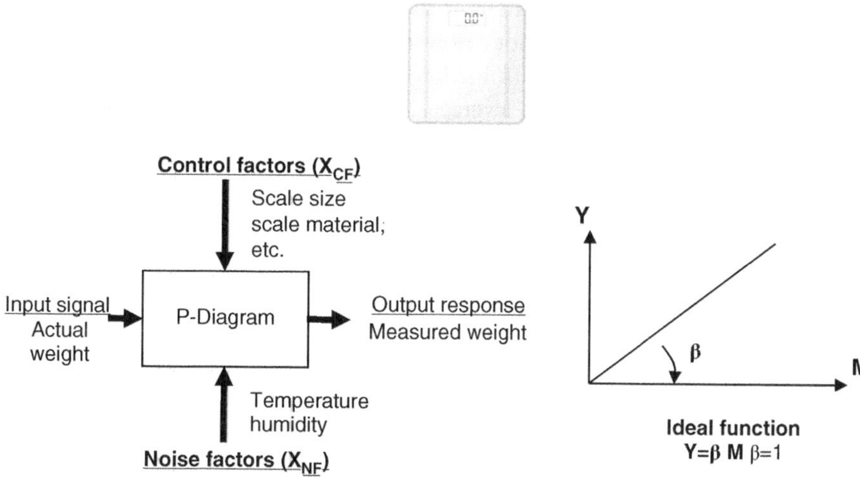

Figure 3.2 Weight Scale P-Diagram.

Ideally, the resulting weight measurement should be equal to the actual weight. Thus, the ideal function here is a straight line with a slope of one in the signal–response graph. This relationship must hold for all operating conditions. However, the presence of noise factors causes the relationship to deviate from the ideal.

The parameter diagram, or P-diagram, is a useful tool in brainstorming and documenting:

- Ideal functions.
- Noise factors.
- Control factors, input signal, (s) and
- Error states (or the failure modes).

The P-diagram describes a system to be engineered. Essentially, it is a schematic diagram that encompasses control factors, noise factors, signal factors, and ideal functions, as shown in Figure 3.3.

A product can be divided into functionally oriented operating systems. "Function" is a key term and a fundamental aspect for describing our product's behavior. Regardless of the

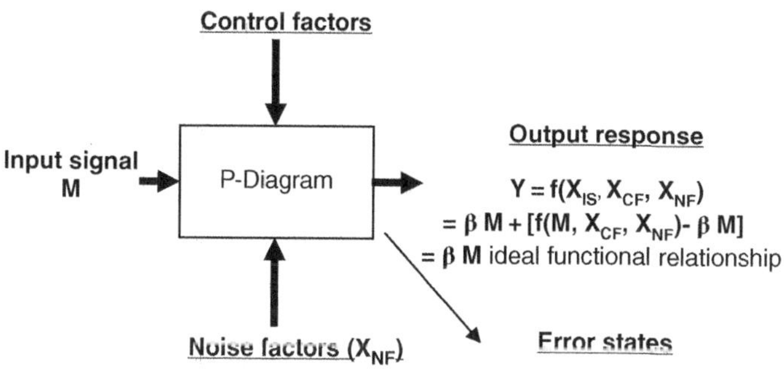

Figure 3.3 P-Diagram.

method used to facilitate a design, they all must start with an understanding of functions. However, what is the definition of function? How is function defined in these disciplines? Understanding the specific meanings of function (or the definition of function) within each of these disciplines could help us leverage tools to improve design efficiency and effectiveness.

According to Webster dictionary, function has three basic explanations as follows:

- the acts or operations expected of a person or thing,
- the action for which a person or thing is specially fitted or used, or
- to operate in the proper or expected manner.

Generally, people would agree that a function describes what a thing does and can be expressed as the combination of a noun and a verb, such as creating a seal or sending an email. The delivery of a function is usually accomplished by one or several transformations. For instance, the function "send an email" is primarily accomplished through "signal transformation." On the other hand, the function "accelerate automobile" is primarily achieved through "energy transformation" (conversion of chemical energy in gasoline to mechanical energy), along with some material and signal transformations (engine control). There is always a main transformation associated with a function. In all technical systems, these main transformations are controlled transformations. For example, in the function "accelerate automobile," the main transformation "energy transformation" is a controlled one, and it is regulated by the driver through the gas pedal.

Main function is a noun–verb description of the function of a system that is related to the main task of the system.

Main transformation is the major type of transformation that is related to the main function.

The ideal function is a description of the main transformation expressed in mathematical form, such as $Y = \beta M$, where M is an input signal (IS) often related to the user's intent and the level of control. Y represents an output response (OR), often indicating the magnitude of the main transformation as shown in Figure 3.4.

Input signal is an input to a system provided by the user, which carries the user's intent for Sensitivity **(β)** is the slope of the least squares best fit line for response data.

Control factors are design choices as shown in Figure 3.5 that influence a response or its variability. They represent product or process design parameters for which the engineer can freely specify a nominal level.

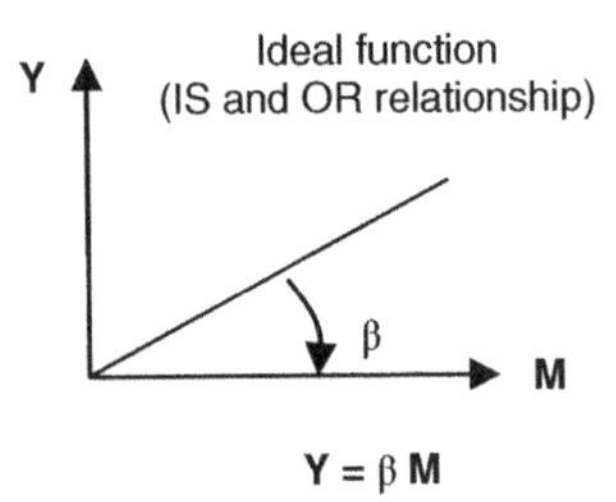

Figure 3.4 Ideal Function (Input–Output Relationship).

Figure 3.5 Effect of Control Factor Levels on Response Mean.

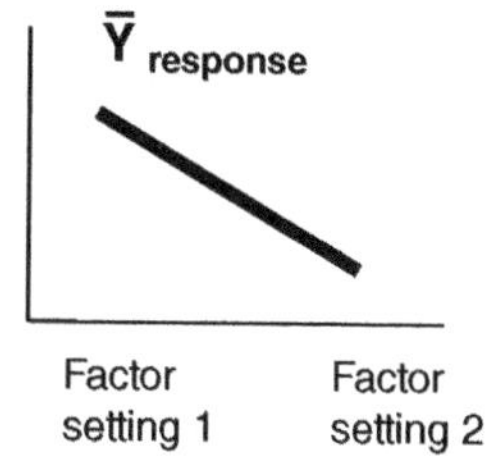

Figure 3.6 Example of Scaling Factor Effect on Mean Response.

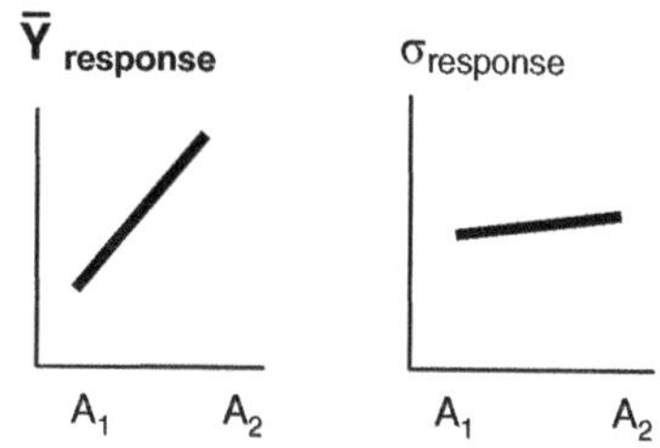

Scaling factors are control factors as shown in Figure 3.6 that have a significant effect on the mean response but little impact on the variability in the response. Use a scaling factor to shift the mean toward its target while potentially incurring a small loss in robustness.

Economic factors are control factors as shown in Figure 3.7 that have little effect on either the variability in the response or the mean response. These factors have minimal impact on robustness improvement or the adjustment of the means.

Noise factors are customer usage or manufacturing process conditions that the engineer either cannot control or decides not to control. These factors are selected for their ability to impact the transfer functions. The main purpose of noise is to introduce variability in the output response. Noise factors can be categorized as piece-to-piece variation, interactions or coupling with other components, subsystems, or the entire system, customer usage, environmental factors, and degradation [4] (Hu, 2013).

When the control factor A interacts with the noise factor N (refer to Figure 3.8), the nominal value of A can significantly impact response variability. Adjust the level of this control factor to desensitize performance to the noise factor N. Level A_2 is a robust choice to ensure the performance is insensitive to noise.

Should noise factors be eliminated? It depends. Can aging or deterioration effects be eliminated? No! Ensure robustness at the lowest cost. Learn to coexist with the noise factors. Design a system that is robust and insensitive to noise factors.

Figure 3.7 Example of Economic Factor Effect on Mean Response.

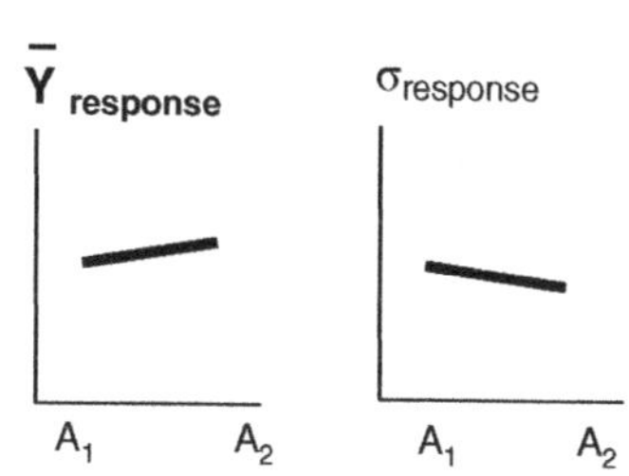

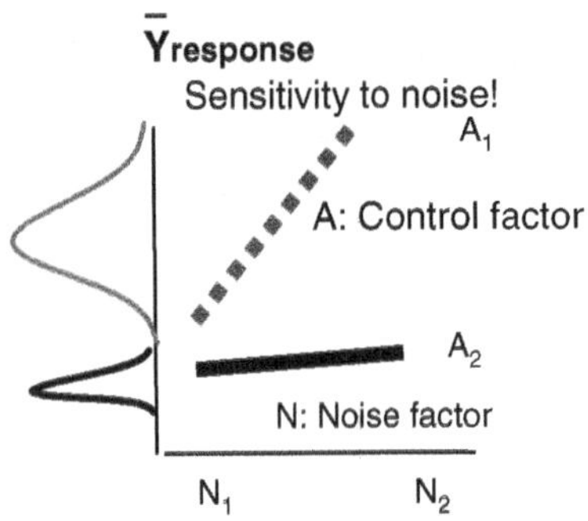

Figure 3.8 Interaction Between Control Factor and Noise Factor.

Error states represent harmful side effects of noise factors and the undesirable output of the system caused by variations, degradation, or the loss of ideal function.

Minimizing error states does not necessarily maximize the design functionalities. However, maximizing the design functionalities will automatically minimize the error states.

The P-diagram is founded on the concept of converting 100% of input energy (input signal) into 100% of ideal function thinking. While every engineered system ideally reaches its "ideal function," achieving 100% efficiency in energy transformation is not realistic. Losses occur, leading to unintended functions or error states. The P-diagram facilitates discussions on countermeasures for failure modes through brainstorming and analysis. For example, in a water pump seal design, leakage is a failure mode and a symptom. Addressing this, one natural consideration is how to stop leakage. The question arises: How can we determine if leakage has stopped or not? The answer lies in observing and evaluating for "no leakage" or "minimum leakage."

Such thinking represents the thought process from the customer's perspective and may not be necessary from an engineering standpoint. From an engineering standpoint, seal design is focused on creating the seal function. A P-diagram for seal design is illustrated in Figure 3.9.

Examining the P-diagram for seal design, we can discern that the true design intent is to provide seal force rather than leakage. Between the input (pressure) and output (seal force), it is a challenging task to minimize the sensitivity β so that the output seal force can

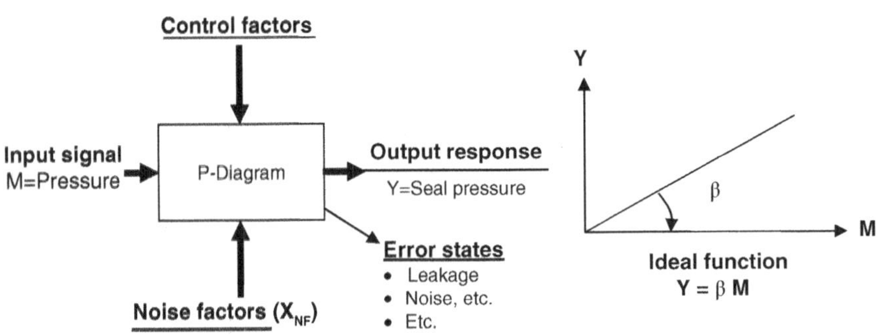

Figure 3.9 Seal Design P-Diagram.

be more stable. Therefore, maximizing the seal force in the presence of uncontrollable user conditions, such as temperature and vibration, is an important engineering task, aligning with the intent of developing a transfer function to optimize the design [4] (Hu, 2013).

The development of a transfer function in DFSS, such as a seal function transfer function for a water pump design, is likely one of the most creative and challenging tasks. This holds true not only due to the complexity involving many field variables in the process, tools, and material properties but also in selecting the proper system output response. A well-developed transfer function can be highly beneficial in assessing the gap between conceptual design and parameter design, emphasizing upfront robustness thinking and testability. Refer to the transfer function chapter for details [4, 5] (Hu, 2013).

The designer's task is to choose appropriate control factors and their settings to minimize the deviation from the ideal at a low cost. Such a design is termed a minimum sensitivity design or a robust design. Achieving this involves exploiting the nonlinearity of the products/systems. The robust design method prescribes a systematic procedure for minimizing design sensitivity, known as parameter design.

An overwhelming majority of product failures and the resulting field costs and design iterations result from neglecting noise factors during the early design stages. The noise factors emerge one by one as surprises in the subsequent product delivery stages, leading to costly failures and the need for band-aids. The robust design method avoids these problems by subjecting design ideas to noise factors through parameter design.

The next step is to specify the allowed deviation of the parameters from the nominal values, involving a balance between the added cost of tighter tolerances and the benefits to the customer. Similar decisions must be made regarding the selection of different grades of subsystems and components from available alternatives. The quadratic loss function is highly useful for quantifying the impact of these decisions on customers or higher-level systems. The process of balancing costs is called tolerance design.

3.5 Opportunity for Robustness Development

Interactions between control factors and noise factors provide opportunities for robustness improvements. These interactions are utilized to enhance the S/N ratio by employing inner and outer orthogonal arrays. The experimental design strategy involves recognizing controllable factors and noise factors and treating them separately. Taguchi identifies three varieties of noises, as illustrated in Figure 3.10, representing potential sources of noise. A product or process may be sensitive to all three forms of noise simultaneously [2, 3].

Taguchi proposes an inner and outer orthogonal parameter design that segregates controllable factors (inner array) from uncontrollable factors (outer array), as shown in Figure 3.11. This parameter design strategy is more robust compared to simply using repetitions to assess noise effects. Taguchi assigns noise factors to the outer array to identify control factor levels in the inner array that produce low variation in the product or process despite the presence of noise factors [2, 3].

Taking advantage of noise factors is one of the key tasks in robust design. The concept of an inner array and outer array is illustrated in Figure 3.11.

Deviation of a functional characteristic from target value	Sources of noise		
	Between unit	Outer noise	Inner noise
	Piece-to-Piece Mfg. & supply chain variation	Customer usage, operating environment & variation from other systems	Wear, Aging, Deterioration

Figure 3.10 Sources of Noise.

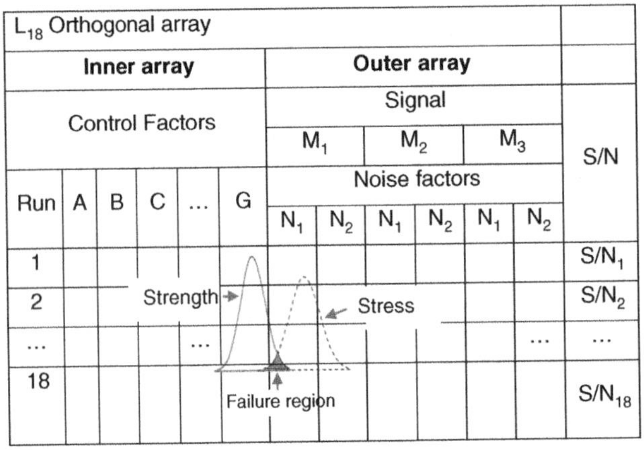

Figure 3.11 Inner Array and Outer Array.

Factor to Reduce Variability:

- When the control factor A interacts with the noise, the nominal value of A can significantly impact response variability.
- Adjust the level of this control factor to desensitize performance to the noise and decrease response variability. As shown in Figure 3.12, the level of A_2 is not sensitive to the noise factor. We can consider selecting the A_2 to minimize variability.

Factor to Adjust Mean Value: When control factor A does not interact with noise factor N (Figure 3.13), changes in noise produce the same response variation regardless of the level of A, as shown by the parallel lines. In this case, A affects only the mean response but not its variability, which is entirely governed by the noise factor.

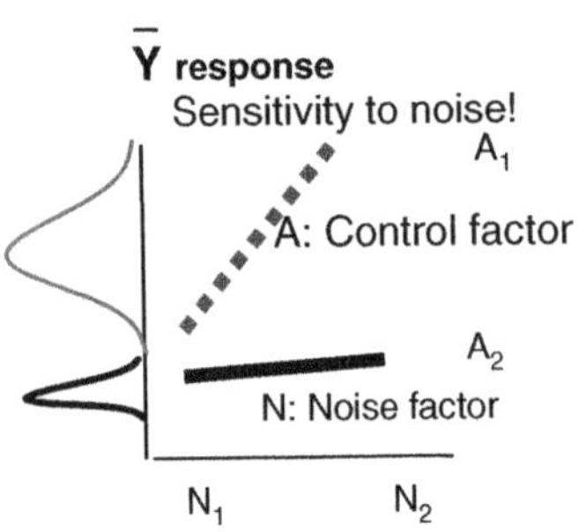

Figure 3.12 Control Factor Interaction Demonstrating Sensitivity to Noise.

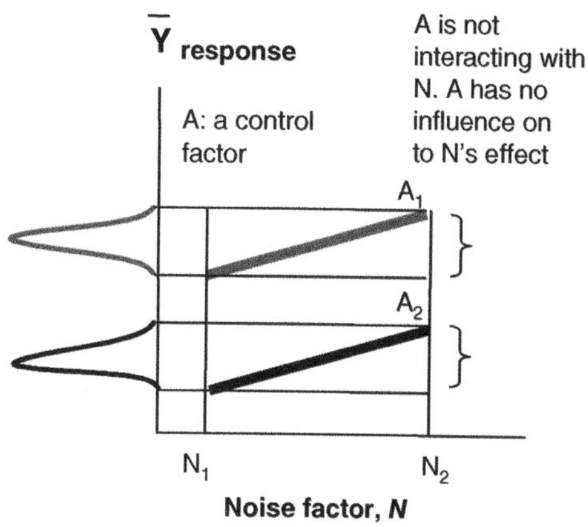

Figure 3.13 Control Factor with No Interaction with Noise.

- When the control factor x does not interact with the noise, the nominal value of A has no effect on response variability.
- Change the level of this control factor to shift the response without affecting variability.

3.6 Two-Step Optimization

There are two common paradigms among design engineers: one involves attempting to hit the target in the first place, and the other centers around improving quality through the use of higher quality (and higher cost) raw materials or components. Taguchi's approach represents a new paradigm—he never tries to hit the target the first time. The first step in quality engineering is to enhance and maximize stability or robustness. Adjusting the average is the second step. In engineering, reducing variability is more challenging than adjusting the mean. While achieving a system's desired target response is crucial, improving robustness requires minimizing the variability of that response. As mentioned earlier, reducing variability is the primary goal of robust engineering. So, how is this accomplished? [2, 3]

It is accomplished by using two-step optimization:

- Step 1: Reduce the variability of each response around its current mean value (i.e., minimize the inherent variation in each response characteristic).
- Step 2: Shift the mean value of the response to the target value (i.e., minimize off-centering).

Two-Step Optimization Procedure:

- Use the experiment to identify control factor settings that maximize S/N ratio.
- Identify a control factor that does not affect S/N ratio and has influence of mean value; use that factor to adjust the mean to target if possible.

This is a different strategy than the traditional product development strategy of getting the target response first. See the dart game example below. Two-step optimization is recommended because it is better to first design a system to be inherently less variable than to first design the system to achieve target response.

Here is the rationale. The engineer first identifies all possible design factors affecting variability. For each factor, the engineer specifies a nominal level that will minimize system performance variability under customer usage. The second step, adjusting the system performance to a particular target level, is easier to achieve. The adjustment to target will be done after minimizing performance variability.

Engineers who are knowledgeable about the systems for which they are responsible are usually aware of how to adjust the response to target as long as the response is an energy-related response. The archer knows that after shooting all the arrows into a tight cluster on the target, one can adjust the bow position or stance to consistently hit the bull's eye. The heat treat engineer knows a shaft's depth of hardness can be increased by increasing cure time.

The more difficult task is to reduce response variability. Once response variability is reduced in the product or process, adjusting the mean response to target is relatively easy.

Two-step optimization plays a pivotal role in robust engineering. In the following section, our focus will be dedicated to discussing the development of robust technology.

3.7 Robustness Measurement: S/N Ratio

The S/N ratio is the mathematical formula employed to calculate design robustness. The larger the S/N ratio, the more robust the performance. This ratio provides insight into how closely the design approaches the optimum performance of a product or process.

It has been utilized in the communications industry to compare a line signal (representing the desired outcome) with line noises (indicating the undesired outcome).

Dr. Taguchi emphasizes that variability in product/process function is the enemy of quality, relating this deviation from the ideal to a concept called noise. Noise, in essence, is variability. Furthermore, factors causing variations are termed "noise factors," which are, by definition, uncontrolled either due to practical limitations or cost considerations [2, 3].

Now let us use Figure 3.14 to discuss how to evaluate systems based on robust engineering perspective.

Considering the S/N ratio mentioned earlier, when we prioritize the intended function and manage to increase the S/N ratio of a system, we can be confident that the system is channeling more energy toward its intended function. Consequently, the overall energy directed toward symptoms of poor quality, including poor reliability, is reduced, leading to a decrease in problems.

On the contrary, let us consider a scenario where one of the unintended functions is measured. If the energy directed toward the corresponding metric is effectively reduced, we cannot guarantee that the redirected energy will be allocated to the intended function and not to another unintended function. The elimination of one unintended function may only lead to the creation of another. This is why robust engineering, and parameter design in particular, emphasizes the measurement of the intended function. The inherent advantage of this approach is engineering efficiency—effectively extracting the most pertinent technical information from experimentation. This concept, along with the S/N ratio, will be explored more comprehensively later in the course. Remember, in parameter design, our goal is to

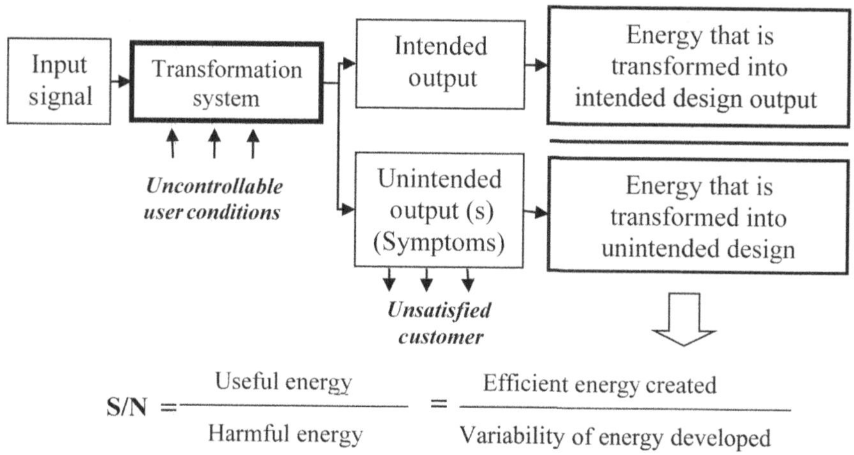

Figure 3.14 Signal-to-Noise Ratio.

achieve a design where S/N is maximized, enhancing efficiency and minimizing variability in energy transformation.

The format of the S/N ratio was developed because Taguchi aimed to optimize the product characteristic for both the location and the variation component. A "transform" was devised for this purpose, known as the S/N ratio. This ratio represents the variation around the target value. The larger the S/N ratio, the smaller the variation. By understanding the S/N ratios of the samples before and after the experiment, Taguchi's loss function can be employed to estimate the potential cost savings resulting from the improved design [2, 3].

The S/N ratio is termed a concurrent statistic, a type of statistic capable of examining two characteristics of a distribution and amalgamating them into a single number or figure of merit. A well-known concurrent statistic used in "classical" statistical analysis is the coefficient of variation (COV).

$$\mathrm{COV} = \frac{s}{\bar{x}} \tag{3.3}$$

One manifestation of Taguchi's S/N figure of merit seems to be the inverse of COV. While this version of the S/N is the least powerful among variable data S/Ns, it is frequently employed to introduce the concept of S/N as a figure of merit. In its basic form, it is the signal (average) divided by noise [2, 3].

$$\mathrm{S/N} = -10\log_{10}\left(\frac{\bar{y}^2}{S^2}\right) \tag{3.4}$$

Here, S represents the standard deviation, and y denotes the average. This S/N ratio is referred to as the nominal-the-best (NTB) case.

The S/N ratios are derived from Taguchi's loss function, and several versions can be constructed to focus on particular aspects of variation and performance. These are the S/N ratios [2, 3].

$$L = K\sum_{\mathrm{i}}(y - \mathrm{target})^2 = K(\mathrm{mean\ shift}^2 + \sigma^2) \tag{3.5}$$

Regardless of what type of S/N ratio, there are always three special components in the form of an S/N ratio (3.6):

$$S/N = -\mathbf{10}\log_{10}\left(\frac{\bar{y}^2}{s^2}\right) \tag{3.6}$$

1) The constant factor (−10) is used to convert the problem from minimizing to maximizing.
2) The log() function ensures that second-order interactions become additive.
3) There is always some form of summation over multiple stress conditions (noise variable arrangements).

Remember that energy is additive in nature. The Log function promotes additivity in the control factor S/N values, allowing them to be summed within the prediction model.

$$S/N_{pred.} = \overline{\frac{S}{N}} + \left(\frac{S}{N_A} - \overline{\frac{S}{N}}\right) + \left(\frac{S}{N_B} - \overline{\frac{S}{N}}\right) + \ldots + \left(\frac{S}{N_P} - \overline{\frac{S}{N}}\right) \tag{3.7}$$

where $S/N_{pred.}$ is the predicted improvement of S/N.
S/N_A is the S/N value of factor *A*.
S/N_B is the S/N value of factor *B*.
S/N_P is the S/N value of factor *P*.

The **Log(•) Transformation:** In the presence of severe, moderate, or mild interactive effects (co-dependencies) between control factors, such as factors A and B, the log transform is employed to promote additivity, as indicated by the following relationship:

$$\log A \times B = \log A + \log B \tag{3.8}$$

One of the tangible benefits of applying the log transform is the enhanced ability to recognize repeatability in robustness confirmation experiments.

3.8 S/N Ratio Improvement and Variation Reduction

Suppose we see the following 3 dBs S/N improvement from an experiment. S/N_{BL} is a base line (BL) value before improvement. S/N_{OP} is the optimal performance (OP) value after improvement.

Then,

$$S/N_{OP}\text{-}S/N_{BL} = 3\text{ dB} \tag{3.9}$$

- $3 = 10[\log(\mu_{OP}^2/\sigma_{OP}^2) - \log(\mu_{BL}^2/\sigma_{BL}^2)]$
- $0.3 = [\log \mu_{OP}^2 - \log \sigma_{OP}^2] - [\log \mu_{BL}^2 - \log \sigma_{BL}^2]$
- We set $\mu_{OP} = \mu_{BL}$ due to the ability to adjust the mean on to the target...
- $0.3 = -\log \sigma_{OP}^2 + \log \sigma_{BL}^2$ or...$(\log \sigma_{BL}^2 - \log \sigma_{OP}^2)$
- $0.3 = \log(\sigma_{BL}^2/\sigma_{OP}^2)$ now take the anti-log
- $10^{0.3} = (\sigma_{BL}^2/\sigma_{BP}^2)$
- So we have approximately $2 = \sigma_{BL}^2/\sigma_{OP}^2$
- $2 = \sigma_{BL}^2/\sigma_{OP}^2$ and then $\boldsymbol{\sigma_{BL}^2 = 2\,\sigma_{OP}^2}$

Table 3.1 Variance Improvement vs. S/N Ratio Increases.

Change in S/N	0.3	0.5	1	2	3	6
Reduction in σ^2	1.07	1.12	1.26	1.59	2	4
Percent Change	7.2	12.2	25.9	58.5	100	400

Each S/N improvement of **3 dB** results in the loss decreasing by half. Here we see that the difference in 3 dB between the two S/N ratios works out to provide a ***twice decrease*** in the ***variance.***

There is a full *100% improvement* in Robustness if the initial variance is *twice as big* (2x) as the final variance. The above Table 3.1 provides a better feel for how dBs relate to our interpretation in S/N analysis:

The third row is the *percent change* from the first response to the second, improved response (*column by column*).

3.9 S/N Ratio, the Additive Model, and the Conservative Laws of Physics

One must balance selection of an S/N ratio and critical functional response with three essential facts.

i) S/N ratios do not measure complex functional relationships like that of a high-order first principles math model. They measure the grouped affect the control factors have on the insensitivity of the system output response to the presence of stressful sources of realistic noise.
ii) S/N ratio contributions are summed up in an additive model to predict an overall S/N gain relative to a baseline or benchmark design. The focus is on quantifying a trend in insensitivity. See prediction model below.

$$S/N_{\text{pred.}} = \overline{\frac{S}{N}} + \left(\frac{S}{N_A} - \overline{\frac{S}{N}}\right) + \left(\frac{S}{N_B} - \overline{\frac{S}{N}}\right) + \cdots + \left(\frac{S}{N_n} - \overline{\frac{S}{N}}\right) \tag{3.10}$$

iii) S/N ratios must be matched with output responses that are leading indicators of the contribution of control factors that contribute to the efficient transformation and flow of energy and mass with respect to the interactions with noise factors.

There are two distinct forms of S/N ratios. They are as follows:

- **Static Forms:** There is only one fixed target on the system output response. Used for a single development project. For example:

$$S/N = \frac{\text{Useful Energy Output}}{\text{Harmful Energy Output}} = 10\log_{10}\frac{\text{Mean}}{\text{Variability}} \text{(For Nominal-the-best)} \tag{3.11}$$

Table 3.2 Dynamic and Non-dynamic Signal-to-Noise Ratios.

Static (Non-dynamic) output responses have a single fixed target value including	Dynamic output responses have multiple input signals and the related output responses
• Nominal-the-Best • Smaller-the-Best • Larger-the-Best • Classified Attribute • Operating Window	• Zero-Point Proportional • Reference-Point Proportional

- **Dynamic Forms:** The targets on the system output response depend upon a signal factor. Used for complex system output responses and in developing new technologies (Platforms).

$$S/N = \frac{\text{Useful Energy Output}}{\text{Harmful Energy Output}} = 10\log_{10}\frac{\text{Efficiency}}{\text{Variability}}\text{(For Dynamic)} \tag{3.12}$$

Types of Responses: Non-dynamic and Dynamic

Descriptions, examples, and S/N ratios for the following response types are discussed in the remainder of this chapter. The S/N ratio equation is provided for each response type.

Table 3.2 summarizes response types for both dynamic and non-dynamic.

3.10 The Static Signal-to-Noise Ratios

Static S/N ratio cases are primarily focused on a fixed target as opposed to having the designed in capability of adjustment to changing targets. The three general forms of the S/N ratio will now be studied, and the relationship to the fundamental loss function will be developed.

3.10.1 Nominal-the-Best (NTB) Case

If the response deviates either lower than the target or higher than the target, a two-sided loss function must be utilized. The corresponding S/N transform is called the NTB S/N. Referring to the quadratic loss function defined in Eq. (3.13), the value of m becomes T (target) and the loss function may be written as follows:

$$L(y) = k(y - \text{Target})^2 \tag{3.13}$$

This loss function is double sided and resembles the quadratic loss curve shown in Figure 3.15. If the value of y deviates from the target T in either direction a loss occurs.

The ratio is developed by setting $k=1$; the loss may be generalized by considering the expected value of the loss by summing all the losses for a population and dividing by n items in the population. With these assumptions the quadratic loss function becomes:

$$E(L) = (\Sigma(y - T)^2/n) \tag{3.14}$$

This is referred to as the ST2, the target variation. In Taguchi's methodology, a measure of merit is sought. Since loss is a measure of demerit, Eq. (3.14) must be inverted to yield a

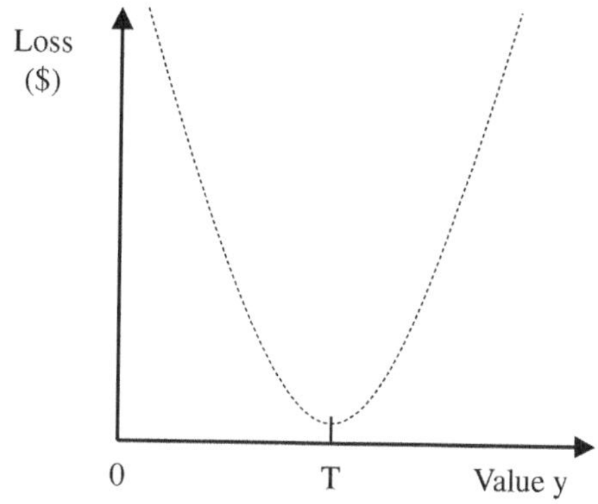

Figure 3.15 Taguchi loss function showing loss versus deviation from target [7].

measure of merit. When inverted, multiplied by the target value, T, and put into Taguchi's log notation, the S/N ratio becomes: [2, 3]

$$S/N_{NTB} = 10 \log_{10} T^2/S_T{}^2 \tag{3.15}$$

More specifically,

$$S/N_{NTB} = 10 \log_{10} \frac{\mu^2}{\sigma^2}$$

Where,

$$\mu = \frac{1}{n}\sum_{i=1}^{n} y_i; \ \sigma^2 = \frac{1}{n-1}\sum_{i=1}^{n}(y_i - \mu)^2$$

A non-negative response with a non-zero target value.
S/N increases as percent variability around the mean decreases.

Examples:

- Coating thickness.
- Dimensions.
- Speed (RPM).

Accordingly, the effect of a factor is equal to the difference between the average S/N ratio for each level, in the case of a two-level experiment, and the difference between the highest and lowest S/N ratios for an experiment with more than two levels. In this analysis, the higher the S/N ratio, the better. In sum, the level for the factor which yields the largest average S/N ratio is to be chosen as the recommended level for that factor.

It is possible to follow the same thought pattern for the remaining S/N expressions.

3.10.2 Smaller-the-Better (STB)

The ratio is developed by using a generalized loss function with k=1, T=0, and the expression (2.14) becomes

$$L(y) = (y - 0)^2 \tag{3.16}$$

Figure 3.16 shows a single-sided quadratic loss function with the minimum loss at zero.

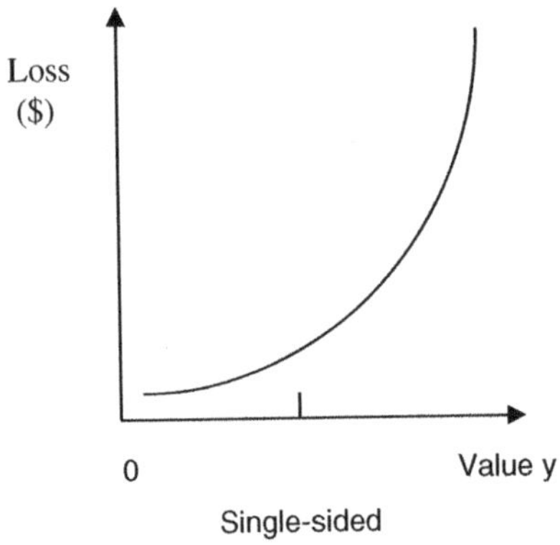

Figure 3.16 STB Quadratic Loss Function.

The loss may be generalized by considering the expected value of the loss by summing all the losses for a population and dividing by n items in the population. With these assumptions the quadratic loss function becomes:

$$E(L) = (\Sigma(y)^2/n) \tag{3.17}$$

This is referred to as the figure of demerit or loss. Taking the negative of this expression a positive quality function or utility function is created.

Taguchi finalizes the S/N ratio by taking the base 10 log and multiplying by 10 to obtain: [2, 3]

$$\mathrm{S/N_{STB}} = -10\log_{10}(\Sigma(y)^2/n) \tag{3.18}$$

$$\mathrm{S/N_{STB}} = -10\log_{10}\left(\frac{1}{n}\sum_{1}^{n} y_i^2\right), y_i \geq 0 \tag{3.19}$$

Specifically, a non-negative response *yi* with a desired value as small as possible, ideally zero. S/N increases as *yi* and σ decreases.

Examples:

- Contamination.
- Leakage.

3.10.3 Larger-the-Better (LTB)

The LTB ratio is defined similarly to the STB S/N, but the y values are squared and then inverted. Figure 3.17 displays the single-sided quadratic loss function. As the response increases, the loss decreases. The quadratic function described by this curve is 1/*y2*.

Figure 3.17 shows that as the values of *y* reach toward infinity, the loss function is minimized. Thus, the implied target for the S/N is infinity or a very large value of the response. The LTB S/N relationship is developed and the final form is

$$\mathrm{S/N_{LTB}} = -10\log_{10}(\Sigma(1/y^2)/n) \tag{3.20}$$

Figure 3.17 LTB Quadratic Loss Function.

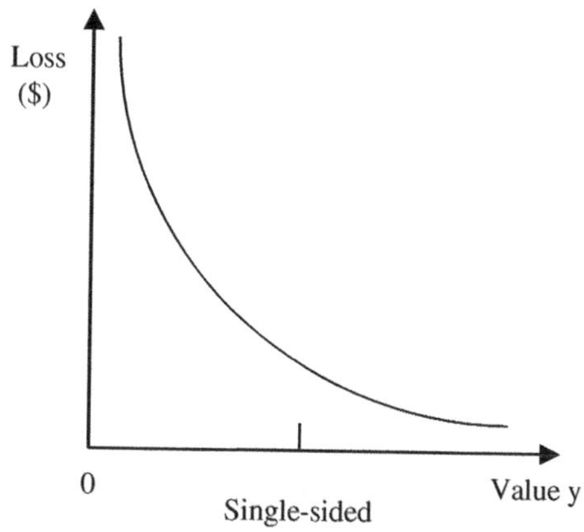

A response, y_i, is as large as possible. S/N increases as y_i increases or σ decreases.

$$S/N_{STB} = -10\log_{10}\left(\frac{1}{n}\sum_{1}^{n}\frac{1}{y_i^2}\right) \tag{3.21}$$

Examples:

- Withstand force.
- Mean time before failure (MTBF).
- Weld strength.

One must keep in mind that since we are finding the inverse of the y values in this calculation, the original responses may not become zero due to the inverse function being undefined. Also, the negative values would lose their meaning. Coding would be the correct procedure to remove negative numbers from the data set.

3.10.4 Operating Window (OW) Response

Operating window (OW) is applicable when a variable has both upper and lower thresholds for good system function. The objective is to maximize the difference between the two thresholds, each of which triggers a different failure mode.

Examples:

- Weld current: Sufficient to create weld but not burn through.
- Spring force on a rotating roller: Sufficient to create single paper feed but not so large as to create multifeed.

This is the most effective method among static characteristic, and the energy level related to the system function is increased gradually from zero. Assume that the energy level when the objective function is generated is x and the energy level when the objective function disappears due to excessive energy is y. Assume further that x is evaluated under n conditions and y is evaluated under k conditions, then the intermediate region between x and y becomes the OW. X is calculated as the smaller-the-better characteristic and y is calculated as larger-the-better characteristic, and their sum is S/N ratio of the OW.

$$S/N_{STBx} = -10\log_{10}\left(\frac{1}{n}\sum_{1}^{n} x_i^2\right), x_i \geq 0 \tag{3.22}$$

$$S/N_{STBy} = -10\log_{10}\left(\frac{1}{p}\sum_{1}^{p} \frac{1}{y_i^2}\right) \tag{3.23}$$

The S/N ratio of the OW is as follows:

$$S/N = S/N_{STBx} + S/N_{STBy} = -10\log_{10}\left(\frac{1}{n}\sum_{1}^{n} x_i^2\right) \times \left(\frac{1}{p}\sum_{1}^{p} \frac{1}{y_i^2}\right) \tag{3.24}$$

3.10.5 Classified Attribute Response

The response is not a continuous variable and performance is classified on a subjective scale. Typically, subjective measures are used because of limitations in measurement technology or cost of measurement.

Examples:

- Appearance: unacceptable, normal, good.
- Color: light red, medium red, dark red.
- Other ratings: pass/fail, good to worst.

Dr. Taguchi recommends classifying such responses into groups, using a minimum of three and a maximum of five groups. Numerical values are then assigned to the groups [2, 3].

Example: After paint, the hood of a vehicle exhibits an "orange peel" appearance. The information is grouped as shown in the graphic at the right. In this case, the data is analyzed using STB.

3.11 Dynamic Signal-to-Noise Ratios

Dynamic S/N ratios are very useful for technology development, which is the process of generating flexible solutions that can be used in many products.

3.11.1 Zero-Point Proportional Response

As previously mentioned, dynamic systems have an input signal, response, and ideal function defined in terms of energy transfer. Let us first review the zero-point proportional case.

Zero-Point Proportional: A system for which, ideally, the response is proportional to the input signal for signal values greater than zero up to the limit of function.

The graphic (as shown in Figure 3.18) at the right illustrates the Zero Point Proportional case. Observe that the line representing Ideal Function passes through the origin.

S/N_{NTB}=$10\log_{10}\frac{\beta^2}{\sigma^2}$

where, $\beta = \frac{1}{r \times r_0}\left(\sum_{i=1}^{n} y_i M_i\right)$;

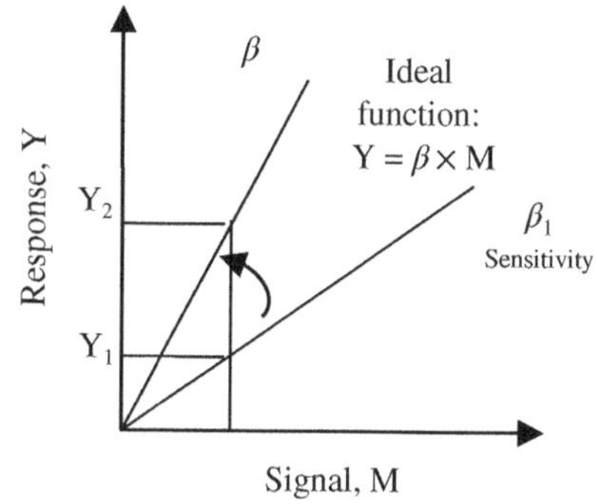

Figure 3.18 Zero-point Proportional Dynamic Response (Ideal Function Through Origin).

$$\sigma^2 = \sqrt{\frac{\sum_{i=1}^{k} \sum_{j}^{r_0} (y_{ij} - \beta \times M_i)^2}{n-1}}$$

n = Total number of response data.
k = Number of levels of signal, M.
r_0 = Number of response data in M.
$r = yi$ is total sum of data under Mi.

Examples:

Product: steering system, pump, hinge, latch, spring.
Process: Machining, stamping, molding, heat treat, paint application.

Reference-Point Proportional Response

There are systems that do not need to have zero output y when input M is zero. Suppose to have a system where output level y_s is expected a certain input Ms, and the ideal relationship between input M and output y is a proportional relationship that goes through point (M_s, y_s) (as shown in Figure 3.19) is called the reference point.

$$y - y_s = \beta(M - M_s) \tag{3.25}$$

Reference-Point Proportional: A system for which, ideally, the response is proportional to the input signal, and it is intentional that function starts at a non-zero signal level. There are two ways of determining y_s.

1) Use average of all data as the standard point M_s.
2) Designate specific data as the standard point M_s. (For example, the first datum or datum from the standard test condition)

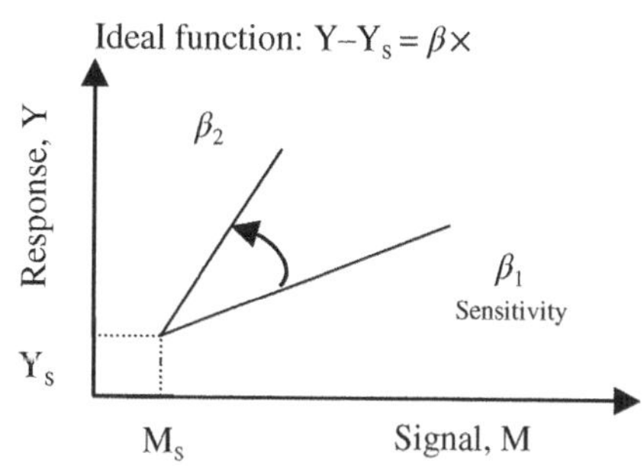

Figure 3.19 Reference-point Proportional Dynamic Response.

After (M_s, y_s) is subtracted from S/N level M and raw data y, the S/N ratio is derived through the same calculation as that for zero-point proportionality.

Let $y' = y - y_s$ and $M' = (M - M_s)$

Equation (2.25) becomes $y' = \beta M'$

The graphic below illustrates this case.

$$S/N_{NTB} = 10\log_{10}\frac{\beta^2}{\sigma^2}$$

Where,

$$\beta = \frac{1}{r \times r_0}\left(\sum_{i=1}^{n} y'_i M'_i\right);$$

$$\sigma^2 = \sqrt{\frac{\sum_{i=1}^{k}\sum_{j}^{r_0}(y'_{ij} - \beta \times M'_i)^2}{n-1}}$$

N = Total number of response data

k = Number of levels of signal, M

r_0 = Number of response data in M,

$r = \sum_{i}^{k} M_i^2$

y_i is total sum of data under M_i

Example:

- Throttle position sensor which functions only above an M_s engine idle condition.
- Thread cutting bolt (clamp load function starts after threads are cut). Torque input to a thread cutting bolt is used to both cut threads and create clamp load. The torque level used to cut threads is the reference torque above which the clamp load function begins. Thus, the response is reference-point proportional.

The selection between a non-dynamic or dynamic formulation often depends on the study's goal, scope, and the practicality of measurement. However, a dynamic formulation is usually preferable because the system is studied at various levels of input. Since data analysis compares how well the data fits a linear ideal function, the analysis is evaluating a system based on the transformation of energy. When a function is nonlinear in nature, appropriate data transformation is performed to linearize the function.

Additionally, all dynamic formulations allow for two-step optimization, whereas non-dynamic formulations, with the exception of NTB, do not permit two-step optimization. Apart from reducing variability, two-step optimization provides information on control factors that can adjust the nominal response to meet the target requirement for a specific product. The knowledge generated is applicable not only to a single product line but also to a family of vehicles or processes.

3.12 Robust Parameter Design Strategy and Steps

Robust parameter design provides an approach to effective designing industrial experiments which will improve the understanding of the relationship between product and process parameters and the desired performance characteristic. This effective parameter design is based on

- Proper system output response selection (What to measure)

It is highly recommended to always select a response measure related to function rather than a symptom of poor function. Measuring symptoms often leads to large control factor interactions that then need to be evaluated in the experiment. If large control factor interactions are not evaluated, the experiment will lead to erroneous conclusions. When designing an experiment to increase robustness, if possible, select a quantitative response measure related to function rather than a symptom of poor function.

Noise Management Strategy

Robust engineering design requires a systematic approach to identifying, understanding, and mitigating the influence of noise factors that disrupt ideal system function. Noise management ensures that product performance remains consistent despite disturbances arising from manufacturing variation, aging, customer usage, environmental conditions, and subsystem interactions. The methodology presented here integrates conceptual models, analytical tools, noise-selection strategies, and verification structures to support robust product development.

Understanding Noise and Its Influence

Noise factors introduce variability that moves the system away from its ideal function. Robust engineering design classifies noise into five categories: piece-to-piece variation, wear-out and drift, customer usage, external environment, and subsystem interactions. Effective noise management quantifies these disturbances and reduces the system's sensitivity to them, thereby increasing the "distance to the failure mode"—a measure of how far system performance remains from unacceptable limits.

Structuring the System with the P-Diagram

The parameter diagram (P-diagram) provides a structured functional overview. It delineates the signal factors driving the system, the control factors available for engineering adjustment, the noise factors that disturb performance, and the resulting ideal and error outputs.

Noise Factor Strategy

The selection of an appropriate **noise factor strategy** is critical in parameter design and robustness evaluation. The purpose of the strategy is not merely to introduce noise, but to introduce **the right amount and type of noise** to allow engineers to discriminate between levels of control factors. Proper noise strategy ensures that the response data contain sufficient separation to reveal how design parameters influence performance under noise.

In practice, this means selecting noise levels that induce **clear, measurable differences** in system response. When noise produces adequate separation, fewer measurements are needed; when noise levels overlap, more extensive sampling is required.

Noise Separation Scenarios

Noise levels N1 and N2 produce distinct response clusters. Figure 3.20 shows an ideal case where two noise levels (N_1 and N_2) produce clearly separated response clusters with no overlap, making the effect of noise easy to identify. While this clear separation helps analyze

Table 3.3 The Five Types of Noise Factors that Disturb Ideal Function.

	Noise Factors		Caused by
Inner Noises (Capacity)	1	Piece-to-piece variation of part properties (such as dimensions)	Production rate
	2	Changes over time/mileage in dimensions or strength (such as wear out, fatigue, chemical degradation	Exposure to repetitive demand
Outer Noises (Demand)	3	Customer usage and duty cycle	Conditions of use
	4	External operating environment	Climatic and road conditions
	5	Internal operating environment (error states from one component being received as a noise factor by another)	Component and system interactions

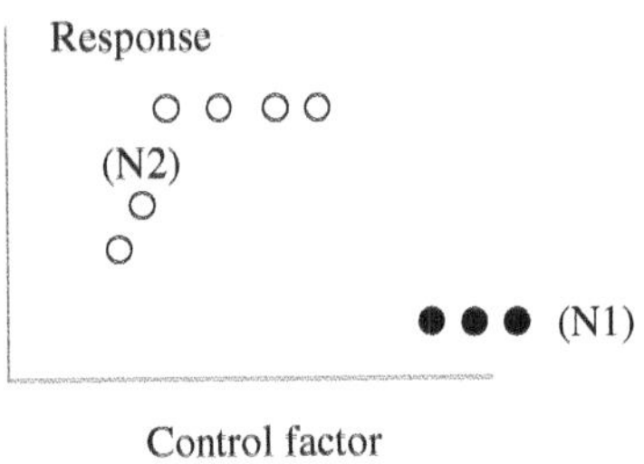

Figure 3.20 Adequate Separation (Ideal Case).

and quantify noise effects, it also indicates that the system is highly sensitive to noise and therefore not robust.

Interpretation:
When separation is clear, *one measurement at each noise level* is typically sufficient. This supports efficient experimentation while preserving analytical resolution.

Noise levels N_1 and N_2 overlap significantly. Figure 3.21 – Insufficient Noise Separation illustrates a situation where the effects of different noise levels (N_1 and N_2) on the system response are not clearly distinguishable.

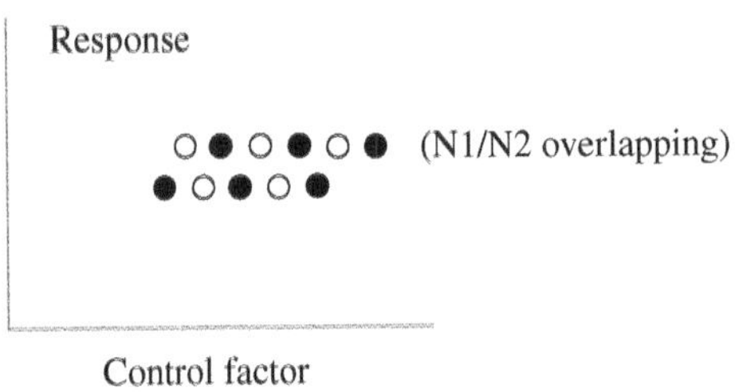

Figure 3.21 Insufficient Noise Separation.

Figure 3.22 Asymmetric or Skewed Noise Effects.

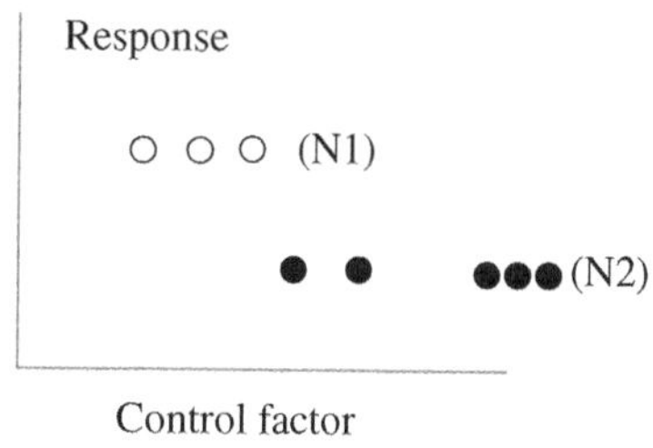

In the figure, the response points corresponding to N_1 (solid dots) and N_2 (open circles) are intermixed and overlapping, forming a single, blended cluster rather than two distinct groups.

Interpretation:
Additional noise levels—or more repeated measurements—are required to distinguish the influence of control factors. Without increased data granularity, robustness conclusions may be misleading or inconclusive.

Noise condition N_2 has greater variability or bias. Figure 3.22 – Asymmetric or Skewed Noise Effects illustrates a situation where different noise conditions do not affect the response equally. Instead, one noise level (e.g., N_2) has a stronger and more variable impact on the response than the other (e.g., N_1).

Interpretation:
When a noise condition (e.g., N2) dominates variability, the engineer must collect *more data under that condition* to understand its influence. For dynamic systems, additional signal levels may also be required.

Guiding Principles for Noise Strategy

- Noise must be **large enough** to reveal differences in sensitivity between control-factor settings.
- The **same noise levels must be applied consistently** across all experimental trials.
- If separation is inadequate, **increase the number of noise levels**, the magnitude of noise, or the number of samples.
- The goal is to achieve **high resolution in response data**, enabling confident estimation of robustness metrics.

This approach ensures that parameter design experiments generate data with sufficient clarity to support optimization of control-factor settings and improvement of robustness.

Selecting the Right Countermeasures

Countermeasures for robustness failures. Countermeasures fall into five strategy types:
A – Change concept/technology
B – Make design insensitive (parameter design)
C – Reduce or remove noise factor
D – Insert compensation device
E – Disguise or redirect noise effects

A complete noise management strategy encompasses not only identification and characterization of noise factors but also **intentional design of the noise factor strategy** itself. Without adequate noise separation, experiments cannot reveal true system sensitivity, and countermeasures may fail. By integrating P-diagrams, noise registers, checklists, Reliability Demonstration Matrix (RDM)s, and noise-strategy principles, engineers create a disciplined, traceable process that leads to inherently robust systems capable of delivering consistent performance under real-world variability.

Table 3.4 Noise–Strategy Matrix.

		A	B1	B2	C	D	E
	Counter measure noise	**Change design concept**	**Parameter design**	**Upgrade**	**Remove noise**	**Compensation device**	**Redirect noise**
1	Part variations	•	•		•		
2	Wear out	•	•		•	•	
3	Customer duty cycle	•	•	•	○		
4	Climate/external environment		•	•	○		
5	Component interactions		•		•	•	•

Note: Parameter design is the only universal strategy.

A fractional factorial experiment-based parameter design allows an experiment to be conducted with only a fraction of all the possible experimental combinations of parameter values. Orthogonal arrays are used to aid in the design of an experiment. The orthogonal array will specify the test cases to conduct the experiment. Frequently, two orthogonal arrays are used: a design control factor matrix and a noise factor matrix, the latter used to conduct the experiment is the presence of difficult to control variation so as to develop a robust design. This approach to designing and conducting an experiment to determine the effect of design factors (parameters) and noise factors on a performance characteristic is represented in Figure 3.23.

These experimental results can be summarized into a metric called the S/N ratio which jointly considers how effectively the mean value (signal) of the parameter has been achieved and the amount of variability that has been experienced. As a result, a designer can identify the parameters that will have the greatest effect on the achievement of a product's performance characteristic.

The design parameters or control factors are identified in an inner array or design factor matrix which specifies the factor level or design parameter test cases.

Control Factors and Levels. As discussed earlier in this chapter, control factors are inherent to a specific product or process and are specified by the engineer. Their nominal levels are established in the design/development process.

Control factors are included in an experiment at more than one level. The purpose for including two or more levels is to identify the level that best minimizes response variability in the presence of noise. Minimum variability means maximum robustness.

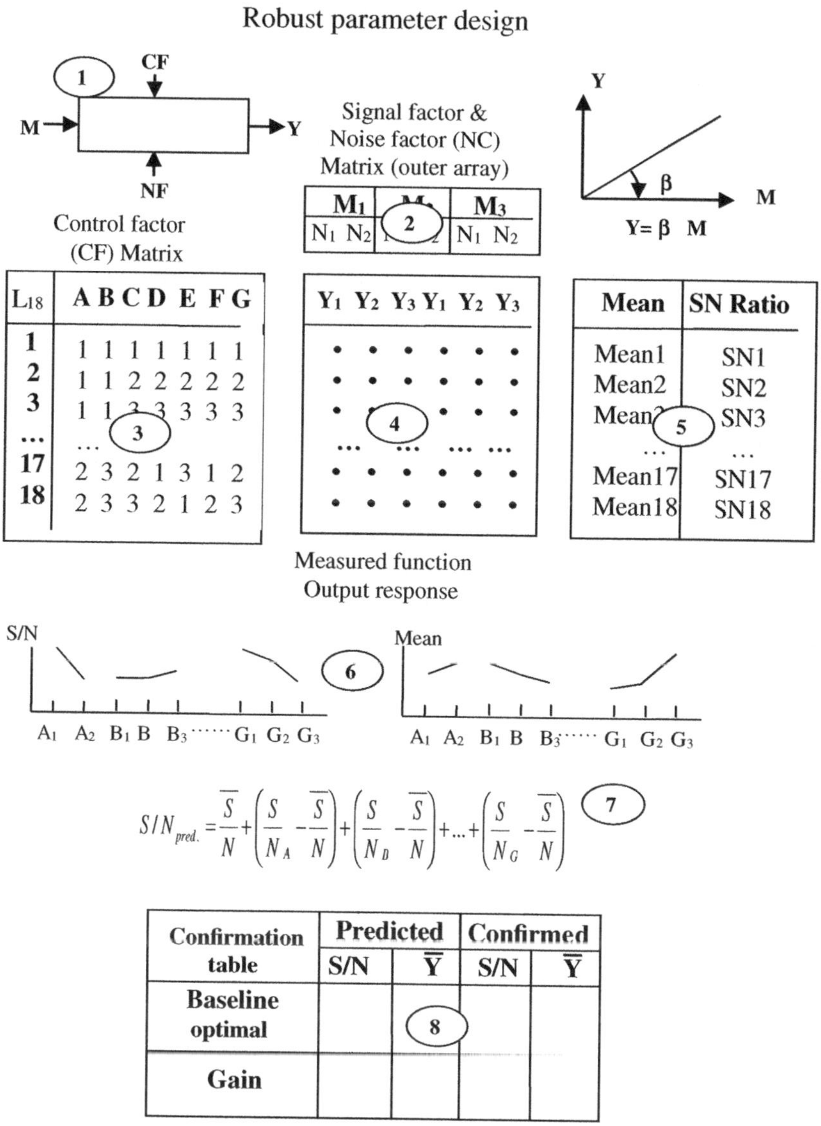

Figure 3.23 Robust Parameter Design Steps.

Table 3.5 Control Factor Examples.

Control Factor Examples	
Product	**Process**
Type of Material	Machining cutter speed
Rivet Diameter	Part holding pressure
Spring Force	Type of cutting tool
Shaft Length	Fixture height

The outer array or noise factor matrix specifies the noise factor or the range of variations that the product will be exposed to in the manufacturing process, the environment, or how the product used (conditions it is exposed to). Data for each control factor configuration is collected at each level of noise. The outer array at the right shows a single noise factor at n levels, N_1, N_2...N_n. Noise factors are factors that cause variability of function and that the engineer either cannot control or chooses not to control, in some cases, because of cost. Noise factors from the customers' environment and aging are usually impossible to control. The purpose for including noise in a parameter design experiment is to induce variability in the response and allow the engineer to select a design exhibiting the least response variability.

This experimental set-up allows the identification of the design parameter values or factor levels that will produce the best performing, most reliable, or most satisfactory product over the expected range of noise factors or environmental conditions.

After the experiments are conducted and the S/N ratio determined for each design factor test case, a mean S/N ratio value is calculated for each design factor level or value. This data is statistically analyzed using analysis of variation (ANOVA) techniques. Very simply, a design factor with a large difference in the S/N ratio from one factor setting to another indicates that the factor or design parameter is a significant contributor to the achievement of the performance characteristic. When there is little difference in the S/N ratio from one factor setting to another, this indicates that the factor is insignificant with respect to the performance characteristic.

With the understanding from the experiments and subsequent analysis, the designer can:

- Identify parameter values which maximize achievement of performance characteristic and minimize the effect of noise, thereby achieving a more robust design.
- Identify parameters that have no significant effect on performance. In these cases, tolerance can be relaxed, and cost reduced.
- Identify parameter values which reduce cost without affecting performance or variation.

These steps take an initial effort but can reduce cost and improve the performance of the product. In the past, the designer selected design parameters and tolerances and made system design trade-offs in an intuitive manner, sometimes supported by limited analysis and trial-and-error experimentation. However, an overall framework was lacking to make these decisions. Design of experiments techniques offer a framework for developing a more rigorous understanding of the relationship between product and process parameters and the achievement of a performance, reliability or quality characteristic, thereby leading to improved design decisions. These techniques present a comprehensive approach experimental design, analysis, and product and process design decision-making.

3.12.1 Steps in Robust Parameter Design for Nominal-the-Best Characteristics

There are eight steps in robust parameter design:

i) Problem Formulation:

This step consists of identifying the main function, developing the P-diagram, defining the ideal function and S/N ratio, and planning the experiments. The experiments involve changing the control, noise and signal factors systematically using orthogonal arrays.

Step 1. Define project scope/objectives:

- Define project objectives.
- Identify the system or subsystems.
- Select team leader and members.
- Establish overall strategies.
- Construct the engineered system (P-diagram). The "engineered system," also known as the P-diagram, identifies signal, control factors, noise factors, and output response of the system. Benefit is to identify and categorize factors that influence the system.
- Identify ideal function/response to be measured. The "ideal function" is a perfect performance of how something works. Benefit is to have paradigm shift. Begins the shift in thinking from looking at quality indicators to looking at how the system functions.
- Establish intent, desired results.
- Define input signal and output response.
- Determine measurement feasibility.

Step 2. Develop and select signal and noise factor strategies. Selecting the "noise factor implementation strategy" involves choosing appropriate noise factors to include in an experiment. Benefit is to identify the noise factors that have the greatest effect on the desired response.

- Define signal levels and ranges.
- Identify all noise factors.
- Select critical noise factors and set levels.
- Determine noise strategies.
- Assign noise factor to outer array.

Step 3. Establish control factors and levels

- Identify all control factors.
- Select critical control factors.
- Select orthogonal array.
- Assign control factors to orthogonal array (inner array).

ii) Conduct Experiments/Data Collection/Simulation:

The experiments may be conducted in hardware or through simulation. It is not necessary to have a full-scale model of the product for the purpose of experimentation. It is sufficient and more desirable to have an essential model of the product that adequately captures the design concept. Thus, the experiments can be done more economically.

Step 4. Plan/prepare for experiment

- Conduct experiment.
- Collect data.

iii) Calculate S/N Ratios and βs and Factor Effects Analysis:

The effects of the control factors are calculated in this step and the results are analyzed to select optimum setting of the control factors.

Step 5. Calculate/interpret response tables

Step 6. Plot factor effect graphs

- Perform two-step optimization.

iv) Prediction/Confirmation:

In order to validate the optimum conditions, we predict the performance of the product design under baseline and optimum settings of the control factors. Then we perform confirmation experiments under these conditions and compare the results with the predictions. If the results of confirmation experiments agree with the predictions, then we implement the results. Otherwise, the above steps must be iterated.

Step 7. Make predictions
Step 8. Conduct confirmation run

- Implement and document results.

3.13 Quality Measurement: Loss Function

In quality improvement and design optimization the metric plays a crucial role. Unfortunately, a single metric does not serve all stages of product delivery. The loss function establishes a financial measure of the user dissatisfaction with a product's performance as it deviates from a target value. Thus, both average performance and variation are critical measures of quality. It is common to use the fraction of products outside the specified limits as the measure of quality. Though it is a good measure of the loss due to scrap, it miserably fails as a predictor of customer satisfaction. The quality loss function serves that purpose very well.

Taguchi believes that the customer becomes increasingly dissatisfied as performance departs farther away from the target. He is interested in two aspects of variation. He is interested in the variation around the average. In addition, he is also interested in understanding the variance as a response itself. Taguchi's parameter design seeks to find the optimal conditions of the input factors that will satisfy the response from both a location and variation criteria [2, 3].

One way of determining this optimal condition is to locate the conditions that will minimize loss. If we were to optimize only on variation, or only on location, we would not have a total quality product or process. To resolve this problem, Taguchi employs the use of a measure that looks at both location and variation simultaneously. Taguchi calls this quality metric the loss function [2, 3].

Taguchi defines quality as "the loss imparted to the society from the time a product is shipped to the customer." The loss is measured in monetary terms and includes all costs in excess of the cost of a perfect product. The conventional method of computing the cost of quality was based on the number of parts rejected and reworked. This method is incapable of distinguishing between two samples, both within the specification limits, but with different distributions of targeted properties. Figure 3.24 displays the conventional method and Taguchi's view of the loss function [2, 3].

Taguchi's loss function is a function of deviation from an ideal target value of the design parameter. T represents the target value or the most desirable value of the parameter under

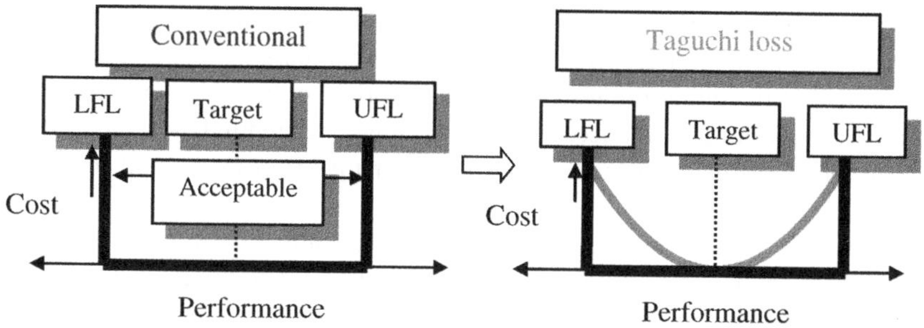

Figure 3.24 Taguchi Loss Function.

consideration. UFL (upper functional limit) and LFL (lower functional limit) represent upper and lower functional limits of a design parameter. It is assumed that if the specified parameter is within this range no loss occurs. However, outside the range total loss is assumed and the product is either discarded or submitted for rework. Taguchi believes that there is no sharp cut-off in the "real" world. Performance begins to gradually deteriorate as the design parameter deviates from its optimum value, T. This is why Taguchi proposes that the loss function be measured by deviation from the ideal value. The loss function is a continuous function and displays the deterioration in product performance when the design parameters deviate from the target value. LFL represents lower consumer tolerance and UFL represents upper consumer tolerance. This is a customer-driven design rather than an engineer's specification. Experts often define consumer tolerance as the performance level where 50% of the consumers are dissatisfied. The following are the key points to remember about Taguchi's loss function [2, 3].

- The cost is smallest when that characteristic is at its target and increases in proportion to the square of the departure from target.
- The "steepness" of the quadratic loss function is determined by the consequences of departing from target.
- A quadratic loss function can be generalized by allowing the cost to have a different shape on one side of the target than on the other side (e.g., adhesive strength of a bonding material).
- A one-sided spec is easily handled with a one-sided quadratic loss function.

Loss can be measured per part. Measuring loss encourages a focus on achieving less variation. As we understand how even a little variation from the nominal results in a loss, the tendency would be to try and keep product and process as close to the nominal value as possible. This is what is so beneficial about the Taguchi loss. It always keeps our focus on the need to continually improve [2, 3].

3.14 Robust Technology Development

Robust technology encompasses systems, products, or technologies engineered for high reliability and resilience, capable of consistent performance under diverse conditions and stressors. Its design prioritizes the ability to endure failures, errors, and external disruptions without substantial impact on functionality. The hallmark of robust technology lies in its adeptness at handling unforeseen events, adapting to dynamic environments, and swiftly recovering from disturbances. Typically, it integrates redundancy, fault tolerance, and error-handling mechanisms to ensure uninterrupted operation and minimal downtime.

Critical sectors like transportation, healthcare, telecommunications, and financial services heavily rely on robust technology. These areas demand unwavering reliability and continuous operation, making robust technology development integral to robust engineering. This developmental process systematically enhances and optimizes technologies, ensuring they remain resilient, stable, and adaptable in the face of variations, uncertainties, and external influences. The overarching objective is to craft technologies that not only meet specified requirements but also demonstrate robust performance across a spectrum of conditions and scenarios.

This approach aligns seamlessly with the principles of robust engineering, which emphasize minimizing variability and maximizing stability throughout the design and development of technological solutions.

Taguchi's approach to robust quality development involves shifting quality engineering upstream. Within the framework of robust engineering, robust technology development aims to create technologies capable of consistent high performance under diverse conditions, thereby minimizing the impact of uncertainties and variations. When comparing Figures 3.25 and 3.26, it becomes evident that employing robust engineering as a corporate strategy offers advantages over the conventional engineering approach [2, 3, 6].

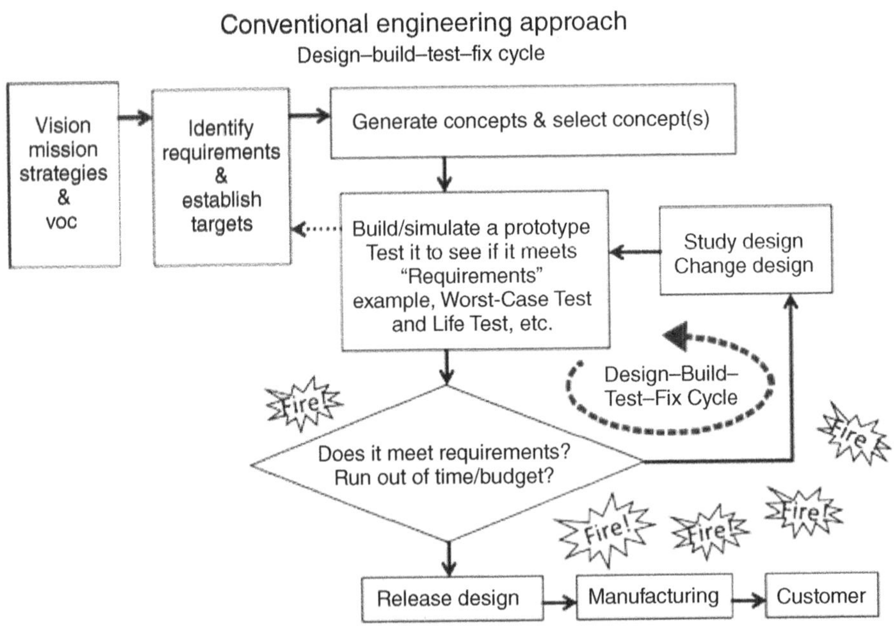

Figure 3.25 Adapted from Taguchi's *Quality Engineering Handbook*.

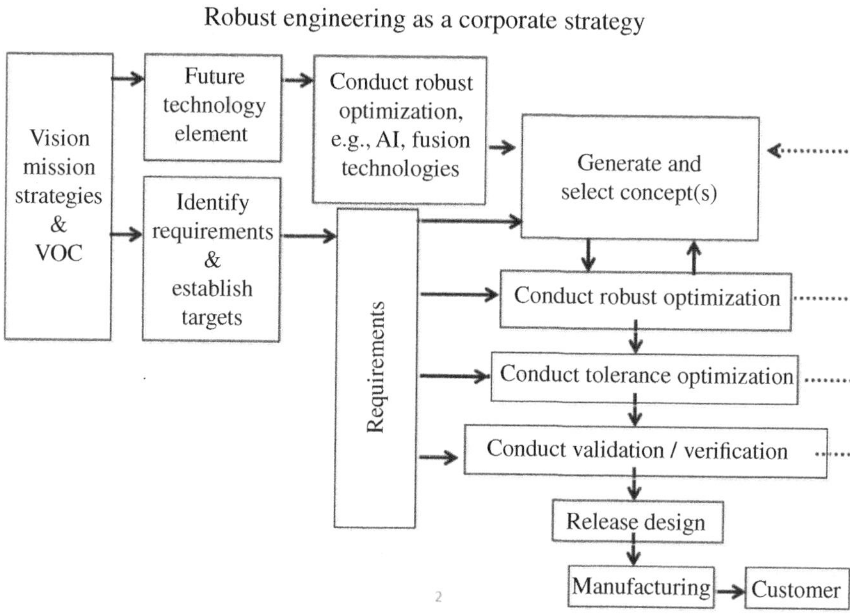

Figure 3.26 Adapted from Taguchi's *Quality Engineering Handbook*.

The "build–test–fix" as shown in Figure 3.25 engineering cycle, also known as the traditional or waterfall development model, has several disadvantages:

Late Identification of Issues:

- **Challenge:** Defects and issues often surface late in the development cycle, typically during the testing phase.
- **Impact:** Identifying problems late can result in higher costs for fixing issues, as changes are more complex and time-consuming to implement.

Limited Flexibility:

- **Challenge:** The model follows a linear and sequential approach, making it less flexible to changes or updates during the development process.
- **Impact:** Adapting to evolving requirements or addressing feedback becomes challenging, leading to a rigid development process.

Longer Time to Market:

- **Challenge:** Sequential phases take time, and the entire cycle needs to be completed before delivering a product to the market.
- **Impact:** Competitors using more agile methodologies may bring products to market faster, potentially affecting market competitiveness.

Customer Feedback Integration:

- **Challenge:** Customer feedback is often incorporated late in the cycle during the testing or post-testing phases.
- **Impact:** Delays in integrating customer feedback can result in products that do not fully meet user expectations or market needs.

Higher Costs of Changes:

- **Challenge:** Making changes after the build phase is expensive and time consuming.
- **Impact:** The cost of changes increases significantly as the development progresses, discouraging modifications and improvements.

Risk of Project Failure:

- **Challenge:** If issues are discovered late in the cycle, there is a higher risk of project failure or delay.
- **Impact:** Late-stage failures can lead to missed deadlines, budget overruns, and potential abandonment of the project.

Less Collaboration:

- **Challenge:** Different teams often work in isolation during specific phases, reducing collaboration.
- **Impact:** Lack of continuous collaboration can result in miscommunication, leading to misunderstandings and misalignments between teams.

Limited Adaptability:

- **Challenge:** The model is less adaptable to changes in requirements or technology during the development process.
- **Impact:** Technological advancements or shifting market needs may not be effectively addressed, leading to outdated solutions.

Quality Assurance Challenges:

- **Challenge:** The testing phase is often constrained by time, leading to potential compromises in the thoroughness of testing.
- **Impact:** Reduced testing can result in undetected defects or quality issues in the final product.

Uncertain Progress Visibility:

- **Challenge:** Limited visibility into progress until the later stages of the cycle.
- **Impact:** Project managers may struggle to assess progress accurately, making it difficult to address potential challenges early on.

As discussed earlier, the "build–test–fix" engineering cycle poses challenges related to late issue identification, inflexibility, longer time to market, limited customer feedback integration, higher costs of changes, risk of project failure, reduced collaboration, limited adaptability, quality assurance challenges, and uncertain progress visibility. These challenges have led to the adoption of more agile and iterative development methodologies in modern software engineering.

Robust engineering, as shown in Figure 3.26, when implemented as a corporate engineering strategy, places a central emphasis on the development of robustness. This approach deliberately steers away from the conventional build–test–fix cycle. Instead of relying on a reactive process of identifying and correcting issues post-production, robust engineering proactively integrates robustness into the design and development stages.

By prioritizing the creation of products or systems that are inherently resilient and less prone to variations, the need for extensive and iterative testing and fixing is reduced. This proactive stance contributes to improved efficiency, reduced development costs, and enhanced predictability in the engineering process.

Some of application examples to illustrate robust technology development:

- **Manufacturing Processes:** Implementing robust technologies in manufacturing processes to reduce the sensitivity to variations in raw materials, environmental conditions, and equipment tolerances. This ensures consistent product quality and minimizes defects.
- **Product Design:** Developing products with robust designs that are less susceptible to variations in user conditions, environmental factors, and manufacturing processes. This leads to increased reliability and customer satisfaction.
- **Automotive Systems:** Designing robust technologies for automotive systems to withstand diverse driving conditions, variations in fuel quality, and environmental factors. This enhances vehicle performance, fuel efficiency, and overall reliability.
- **Communication Networks:** Creating robust technologies for communication networks to ensure reliable and efficient data transmission, even in the presence of signal interference, varying network loads, or unexpected disruptions.
- **Healthcare Devices:** Developing robust technologies for medical devices to operate consistently and accurately, regardless of variations in patient conditions, environmental factors, or user expertise.
- **Energy Systems:** Implementing robust technologies in energy systems to optimize performance under varying loads, environmental conditions, and energy source fluctuations. This improves the overall efficiency and reliability of power generation.
- **Software Development:** Applying robust technology development practices in software engineering to create programs that are less prone to bugs, errors, and variations in user input, resulting in more stable and reliable software applications.
- **Agricultural Practices:** Utilizing robust technologies in precision farming to account for variations in soil conditions, weather patterns, and pest control. This enhances crop yields and overall agricultural productivity.
- **Supply Chain Management:** Implementing robust technologies in supply chain management systems to adapt to fluctuations in demand, changes in logistics, and variations in supplier performance, ensuring a resilient and efficient supply chain.
- **Renewable Energy Systems:** Developing robust technologies for renewable energy systems to harness energy from unpredictable sources like wind and solar, while maintaining stable and continuous power generation.

In each of these examples, robust technology development contributes to increased reliability, performance consistency, and adaptability, aligning with the principles of robust engineering.

Robust design principles play a crucial role in the development of autonomous driving technologies. In the context of autonomous vehicles, robust design aims to create systems that can reliably and safely operate in a wide range of conditions, mitigating uncertainties and variations. Here are some key aspects of robust design in autonomous driving technologies:

Environmental Variability: Autonomous vehicles operate in diverse environments, including varying weather conditions, road types, and traffic scenarios. Robust design ensures that the autonomous system can adapt and perform well under different environmental factors such as rain, snow, fog, and varying light conditions.

Sensor Fusion: Autonomous vehicles rely on a multitude of sensors, including cameras, lidar, radar, and other perception systems. Robust design involves integrating sensor data through sophisticated sensor fusion algorithms, reducing the impact of individual sensor uncertainties, and ensuring reliable perception in complex environments.

Cybersecurity: Autonomous vehicles are vulnerable to cybersecurity threats, including hacking attempts and malicious attacks. Robust design incorporates robust cybersecurity measures to protect the vehicle's systems from unauthorized access and ensure the integrity of critical functions.

Functional Safety: Robust design principles align with functional safety standards, ensuring that autonomous systems have built-in redundancies, fail-safes, and error-detection mechanisms. This is essential for ensuring the safety of passengers and pedestrians in the event of system failures or unexpected situations.

Mapping and Localization: Robust mapping and localization algorithms are critical for autonomous driving. These systems must be designed to handle variations in map data quality, GPS inaccuracies, and changes in the environment to maintain accurate positioning.

Human–Machine Interaction: Robust design extends to the human–machine interface, ensuring that the autonomous vehicle can effectively communicate with passengers and other road users. This includes designing intuitive interfaces and handling communication challenges in various scenarios.

Regulatory Compliance: Autonomous driving technologies need to comply with evolving regulatory standards. Robust design involves staying adaptable to changing regulations and ensuring that the technology can be easily updated to meet new requirements.

Testing and Validation: Robust design includes comprehensive testing and validation procedures. This involves subjecting autonomous systems to a wide range of simulated and real-world scenarios to identify and address potential weaknesses in the design.

Continuous Improvement: Robust design is an ongoing process that involves continuous improvement based on real-world data and feedback. This iterative approach allows developers to address emerging challenges and enhance the system's robustness over time.

Robust design principles are essential for ensuring the reliability, safety, and adaptability of autonomous driving technologies in the face of the complex and dynamic nature of the road environment.

References

1 Matthew Hu, John M. Pieprzak, and John Glowa (2004), Essentials of design robustness in design for six sigma (DFSS) methodology. Appears in the collection Reliability and Robust Design in Automotive Engineering (SP-1844), SAE Technical Paper 2004-01-0813. 13.

2 Taguchi, G., Chowdhury, S., and Yuin, W. (2004). *Taguchi Quality Engineering Handbook*. Hoboken, NJ, USA: John Wiley & Sons.

3 Taguchi, G., Chowdhury, S., and Taguchi, S. (2000). *Robust Engineering, Learn How to Boost Quality While Reducing Costs & Time to Market*. Mc Graw Hill.

4 Matthew Hu and Kai Yang (2005). "Transfer function development in design for six sigma framework," SAE International (Society of Automotive Engineers), Warrendale, Pennsylvania, USA. SAE Technical Paper Number: 2005-01-1418, April 11, 2005, 9.

5 Hu, Matthew, Robustness thinking in design for reliability, RAMS, 2014 *Annual RELIABILITY and MAINTAINABILITY Symposium*.

6 Clausing, D. (1994). *Total Quality Development*. A Step-By-Step Guide to World-Class Concurrent Engineering, ASME Press: New York.

7 Taguchi, G. (1986). Introduction to Quality Engineering: Designing Quality into Products and Processes. Asian Productivity Organization, Tokyo.

4

Robust Design for Reliability (RDfR) A Comprehensive Approach to Product Excellence

CHAPTER MENU

4.1 Introduction

To thrive in today's business landscape, companies must strategically plan their projects with the appropriate level and type of development for maximum efficiency. Research indicates that a significant portion (85%) of issues related to new products, such as functionality problems, delays in market entry, or excessive costs, stem from a subpar design process (Ullman, 1997). A robust design process, as we discussed in Chapter 2, supported by efficient methodologies, is crucial for success.

Recognizing the critical importance of the early phases in the engineering design process for the technical and economic success of a new product, an effective methodology for this stage becomes paramount. Enter robust design for reliability (RDfR), our approach that significantly enhances conventional design for reliability. RDfR serves as an efficient methodology, guiding the "What" and "How To" of engineering product development, with a specific emphasis on reliability to support the introduction of a new product.

A step change in product reliability by improving the quality of our product lifecycle management processes across all engineering, manufacturing centers, and end-to-end supply chain is shown in Figure 4.1 [1].

Moreover, RDfR proves valuable for redesigning and improving aspects of a product, addressing chronic issues effectively. Utilizing statistical tools, RDfR helps comprehend, optimize, and control key factors, ensuring the robust delivery of critical customer attributes in the face of variability. By quantifying risk, RDfR facilitates early business discussions on product delivery quality and reliability during the PDP.

A proactive measure, RDfR plays a pivotal role in the development of prognostic health management (PHM) by enhancing the understanding and management of failure physics. RDfR serves as a foundational approach in PHM development, aligning with reliability design principles, identifying critical parameters, and optimizing system performance amidst variations and uncertainties. This is achieved through:

- Identifying Critical Parameters
- Implementing Design for Reliability
- Developing P-Diagrams
- Considering Noise Factors and Control Factors
- Optimizing Design
- Conducting Reliability Testing and Validation
- Adopting a Proactive Approach
- Pursuing Continuous Improvement

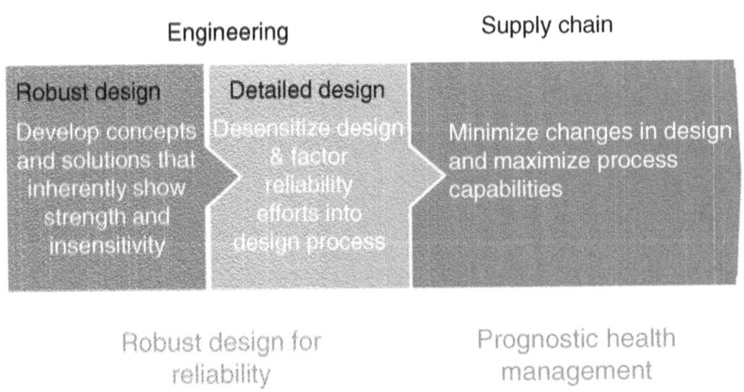

Figure 4.1 RDFR End-to-End Life Cycle Reliability.

This chapter also aims to underscore the significance of identifying and selecting health indicators within the framework of PHM using RDfR principles. By exploring methodologies, best practices, and case studies associated with this critical aspect of PHM, organizations can gain valuable insights to refine their predictive maintenance strategies and enhance reliability and asset performance over time.

We identify noise factors and introduce stresses resulting from these factors to induce failure modes. Following this, we optimize the design under the influence of uncontrollable user conditions, assessing the proximity of ideal performance to potential failure mode(s) and constructing physics-of -failure (PoF) models.

Robust design provides valuable insights for embracing the PoF approach, with a focus on identifying the root cause of failure, conducting comprehensive failure analysis, and understanding failure mechanisms through the analysis of parameter characteristics [2]. This approach entails a specific examination of failure point locations, considering factors such as fabrication technology, process, materials, and circuit layout obtained from the manufacturer. We provide details later to show how our RDfR approach to identifying the need of PHM and taking proactive first step to enable and enhance the prognostic health management (PHM) development in the design stage to optimize the design to ensure the expected performance from the systems or products given certain reliability requirements.

Since most reliability problems stem from variation and uncertainty, designing for reliability requires a direct address of these factors, rather than relying indirectly on failure rates based on often-questionable field data. In contrast to the traditional method of minimizing variation, the principles of robust design focus on desensitizing variation. This ensures that a product or system can consistently perform well under various conditions, all while reducing costs.

The essence of robustness thinking in RDfR lies in making the product insensitive to uncontrollable user environments. In our forthcoming RDfR approach, we will provide a specific definition of uncontrollable user environments. This will enable us to explore strategies for gaining engineering confidence in the DFR during the early phases of product design.

Many companies follow a traditional stage-gate process in which the entire design process is divided into a discrete set of sub-phases (see Figure 4.2).

A typical design process consists of gathering customers' needs, creating design concepts, developing the solution with design steps, testing and validating the design solution, and executing full production. Several variations of these steps exist and sometimes the design step is broken into preliminary design and final design, especially if an iterative approach is utilized. Concept development, preliminary design, and final design are essentially the key stages of design activities.

RDfR is not a replacement for the company's current design process. Instead, it enhances the quality and reliability of the design process and reduces the development cycle time.

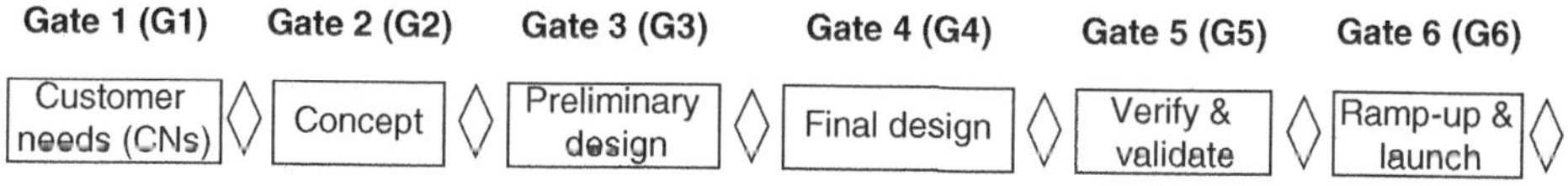

Figure 4.2 Typical Product Development Process (PDP).

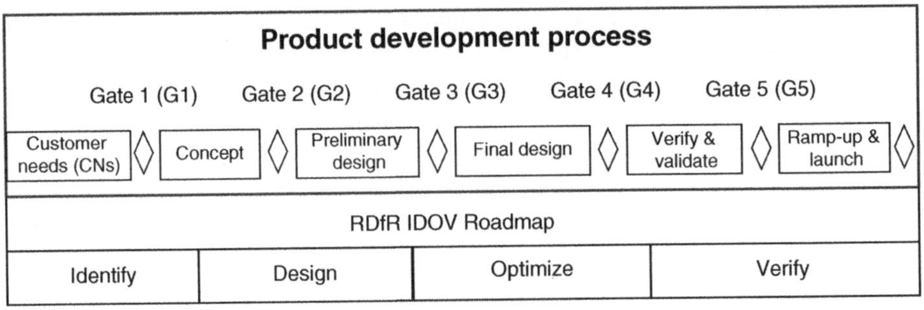

Figure 4.3 DFSS-Enhanced Gate Process.

Neither should it be treated as a set of stand-alone tools, but as a system to be integrated with the existing design process. Toward this end, we must blend RDfR activities and deliverables with the company's new product or service design process. Integration of RDfR activities and deliverables with business or product specific activities and deliverables ensures that the design process is optimized to produce world-class results (see Figure 4.3).

The RDfR approach can utilize any of the many possible methodologies. Each methodology primarily differs in the name of each phase and the number of phases.

The objective of this chapter is to mold RDfR principles, tools, and phases in a comprehensive implementable sequence in a manner that consolidates the advantages of principle-based approaches, and links DFSS phases to the design lifecycle. The lifecycle of a design entity, whether a product, a service, or a process, is usually formed and modeled using phases and tollgates. A "phase" represents a set of design activities of the project and is usually bounded by an entrance tollgate and an exit tollgate. A "TollGate" represents a milestone in the lifecycle and has some formal meaning defined by the development process, cascaded to the design team, and recognized by management and other stakeholders. The lifecycle of a designed entity, whether a product, a service, or a process, starts with some form of idea generation, whether in free invention format or using more disciplined creativity such as those surfaced by R&D departments. This is usually followed by several sequential activities. In the lifecycle, the design process is followed by manufacturing or production activities followed by service.

4.2 Robust Design for Reliability: A Comprehensive Approach to Product Excellence

In today's competitive market, RDfR has emerged as a critical framework for organizations aiming to develop reliable, high-performance products.

The alignment of RDfR with the PDP and making designs insensitive to uncontrollable user environments are crucial aspects of successful RDfR implementation. These strategies are fundamental to creating products that are not only reliable but also resilient in diverse real-world conditions.

Aligning RDfR with PDP ensures that reliability and robustness are not afterthoughts but integral parts of the entire development process. By integrating robustness considerations at every stage, from initial concept to production, companies can proactively address potential issues before they become costly problems.

The development of specific checkpoints and milestones for robustness and reliability creates a structured approach to monitoring and enhancing these critical aspects throughout the product lifecycle.

The creation of cross-functional teams dedicated to robustness and reliability is particularly important [3]. These teams bring together diverse expertise, allowing for a more comprehensive approach to reliability challenges.

This collaborative effort can lead to innovative solutions that might not be apparent when working in silos. Making designs insensitive to uncontrollable user environments is equally critical. Products often face a wide range of usage scenarios that cannot be fully controlled or predicted. By conducting extensive user research, companies can gain valuable insights into how their products will be used in various real-world situations. This understanding is crucial for designing products that can withstand diverse and sometimes unexpected conditions.

Finally, the use of advanced simulation tools to model product behavior in different environments is a powerful strategy [4]. These tools allow companies to predict and address potential issues before physical prototyping, saving time and resources while improving overall product reliability.

By emphasizing these aspects of RDfR, companies can create products that are not only reliable under ideal conditions but also robust enough to maintain their performance in the diverse and often unpredictable environments of real-world use. This approach leads to higher customer satisfaction, reduced warranty claims, and, ultimately, a stronger market position.

This comprehensive approach integrates various aspects of product development to ensure superior reliability and customer satisfaction, while also addressing the challenges of supply chain integration, regulatory compliance, and digital quality.

- Building a Strong RDfR Team
- Concurrent Engineering in RDfR
- Implementation Readiness for RDfR
- Preventing Failure Modes Through Vigilance in RDfR
- Supply Chain Integration in RDfR
- Regulatory Compliance in RDfR
- Digital Quality in RDfR

Building a Strong RDfR Team
Assembling a Cross-functional Team

- Include experts from various disciplines: mechanical, electrical, software, and manufacturing engineering
- Incorporate reliability engineers and data analysts
- Involve quality assurance professionals

Providing the Right Tools and Training

- Equip the team with advanced simulation software and testing equipment
- Implement robust project management systems

- Offer regular training on the latest RDfR technologies and methodologies

Fostering a Culture of Reliability

- Promote robustness and reliability as core company values
- Encourage continuous learning and knowledge sharing
- Implement systems to capture and apply lessons learned from each product cycle

Concurrent Engineering in RDfR
Simultaneous Development

- Develop design and manufacturing processes in parallel
- Identify and address potential issues early in the development cycle

Cross-functional Collaboration

- Facilitate cooperation between different engineering disciplines from the early stages
- Ensure all aspects of product design and manufacturing are considered

Iterative Process

- Implement continuous feedback loops between design and manufacturing teams
- Allow for rapid iterations and improvements

Lifecycle Consideration

- Consider the entire product lifecycle during the design phase
- Ensure robustness and reliability throughout the product's life

Implementation Readiness for RDfR
Aligning RDfR with Product Development Process (PDP)

- Integrate robustness considerations at every stage of the PDP
- Develop specific checkpoints and milestones for robustness and reliability

Making Design Insensitive to Uncontrollable User Environment

- Conduct extensive user research
- Implement Design of Experiments (DoE)
- Utilize advanced simulation tools

Early Development of Robustness

- Implement Failure Mode and Effects Analysis (FMEA) in the concept phase
- Use Quality Function Deployment (QFD)
- Conduct preliminary robustness testing on early prototypes

Setting Reliability Goals Based on Survivability

- Develop detailed survivability models
- Use statistical tools to set realistic reliability targets
- Implement continuous monitoring and adjustment of reliability goals

Maximizing Engineering Confidence

- Conduct extensive simulation and virtual testing

- Implement Design for Six Sigma (DFSS) methodologies
- Develop a knowledge base of robust design principles

Employing Physics of Failure (PoF)

- Develop detailed physical models for potential failure modes
- Integrate PoF analysis with Computer-Aided Engineering (CAE) tools
- Conduct accelerated life testing based on PoF principles

4.2.1 Preventing Failure Modes Through Vigilance

The concept of preventing failure modes through vigilance is a critical component of RDfR [5]. This approach emphasizes the importance of proactive measures to identify, prevent, and mitigate potential failures throughout the product lifecycle. By integrating vigilance into the design and manufacturing processes, organizations can significantly enhance product reliability and performance.

4.2.1.1 Understanding the Entropic Nature of Mistakes

Mistakes and failures represent the natural tendency toward disorder, often referred to as the entropic state. This concept is encapsulated in the adage, "when Murphy speaks, listen," reminding us of the constant need for attention to potential errors and disruptions. In the context of RDfR, this understanding forms the foundation for a proactive approach to reliability.

Entropy in Product Development

- Recognize that systems naturally tend toward disorder over time
- Understand that without constant vigilance, product quality and reliability will degrade
- Implement systems and processes that actively counteract entropic tendencies

Enhancing Individual Vigilance

- Individual vigilance is a crucial factor in preventing failure modes. By fostering a culture that encourages learning from mistakes, organizations can significantly improve their ability to identify and prevent potential failures.

Learning from Mistakes

- Provide opportunities for individuals to recognize and learn from their errors
- Implement a "two-chance" system, allowing individuals to make and correct the same mistake twice
- Use mistake analysis as a learning tool, not a punitive measure

Continuous Improvement Mindset

- Encourage a culture of continuous learning and improvement
- Implement regular training and skill development programs
- Foster an environment where questioning and improvement suggestions are welcomed

4.2.1.2 Strengthening Organizational Vigilance

Organizational vigilance involves implementing systematic approaches to identify and prevent potential failure modes across the entire product development and manufacturing process.

Establishing Operational Definitions (Deming)

- Create clear, concise definitions for all operational processes
- Ensure consistency in understanding across different departments and teams
- Regularly review and update operational definitions to reflect current best practices

The Importance of Operational Definitions

"An operational definition interprets; it is not open to interpretation." – W.E. Deming

Operational definitions are essential tools in ensuring clarity and consistency in communication, particularly within technical and quality management fields. By providing precise and unambiguous criteria, they help prevent misunderstandings and ensure that everyone involved has the same understanding of key concepts and standards.

Why Operational Definitions Matter

1) Clarity in Communication:
   - Operational definitions enable us to know exactly what we are talking about. They provide clear, specific, and measurable criteria for terms and requirements, leaving no room for ambiguity.
2) Preventing Misinterpretation:
   - These definitions help avoid the possibility of misinterpretation of requirements and standards. When everyone understands the exact meaning of a term or standard, the chances of errors and inconsistencies are greatly reduced.

Example of Operational Definitions in Action

Consider the system design specification (SDS) reliability requirement for dynamic seals, which states:

"The dynamic sealing system must meet 0.9989835 reliability for 150,000 miles." This definition raises several important questions that need precise answers:

- What is expected of the sealing system over 150,000 miles in vehicle design?
  - Does it need to maintain a specific level of performance, or is it simply required not to fail?
- How is reliability measured to a seven-digit accuracy?
  - Ensuring such precise measurement requires a clear and standardized method for testing and validating the system's performance.

Without an operational definition, terms like "reliability" and "150,000 miles" could be interpreted in various ways, leading to confusion and inconsistency in implementation and evaluation.

Ensuring Effective Operational Definitions

To create effective operational definitions, consider the following steps:

1) Identify the Key Terms:
   - Determine which terms and standards need clear definitions. Focus on those that are critical to the project's success and are prone to misinterpretation.
2) Provide Detailed Criteria:
   - For each term, provide detailed, measurable criteria that specify exactly what is required. This includes how measurements should be taken, under what conditions, and what standards of accuracy are expected.

3) Test and Validate:
   - Test the definitions with real-world examples to ensure they are practical and can be consistently applied. Gather feedback from stakeholders to refine and improve the definitions.
4) Document and Communicate:
   - Clearly document the operational definitions and communicate them to all relevant parties. Ensure that everyone understands and agrees on the definitions before proceeding.

These definitions are a cornerstone of effective quality management and are crucial for meeting customer expectations and industry standards.

By focusing on building a strong team, implementing concurrent engineering practices, and ensuring readiness for RDfR implementation, organizations can create products that not only meet but exceed customer expectations. This comprehensive approach to RDfR drives innovation, streamlines product development, and significantly enhances product reliability and customer satisfaction.

Moreover, by integrating supply chain management, regulatory compliance, and digital quality initiatives into the RDfR framework, companies can create a holistic approach to product excellence. This expanded strategy not only ensures product reliability but also addresses the complexities of modern global markets, regulatory environments, and technological advancements.

The result is a robust, compliant, and digitally enhanced PDP that can adapt to changing market conditions, meet diverse regulatory requirements, and leverage cutting-edge technologies to maintain a competitive edge. By embracing this comprehensive RDfR approach, organizations can position themselves as leaders in product quality and reliability, driving long-term success and customer loyalty in an increasingly demanding marketplace.

Supply Chain Integration in RDfR
Collaborative Supplier Relationships

- Develop strategic partnerships with key suppliers
- Implement joint reliability improvement initiatives
- Share RDfR best practices and knowledge across the supply chain

Supply Chain Risk Management

- Conduct comprehensive risk assessments of the entire supply chain
- Implement mitigation strategies for identified risks
- Develop contingency plans for potential supply chain disruptions

Supplier Quality Assurance

- Establish clear reliability and quality standards for suppliers
- Implement supplier audits and performance monitoring systems
- Provide training and support to suppliers on RDfR principles

End-to-End Traceabilitys

- Implement technologies for full product traceability throughout the supply chain
- Use data analytics to identify and address potential reliability issues in the supply chain
- Ensure transparency and accountability at every stage of the product lifecycle

Regulatory Compliance in RDfR
Proactive Compliance Management

- Stay informed about evolving regulatory requirements in target markets
- Integrate compliance considerations into the early stages of product development
- Develop a cross-functional team dedicated to regulatory compliance

Compliance Testing and Documentation

- Implement rigorous testing protocols to ensure compliance with all relevant standards
- Maintain comprehensive documentation of compliance-related activities and test results
- Develop a system for managing and updating compliance documentation

Global Regulatory Strategy

- Develop a strategy for addressing varying regulatory requirements across different markets
- Implement design practices that allow for easy adaptation to different regulatory environments
- Engage with regulatory bodies to stay ahead of upcoming changes and contribute to standard development

Compliance Training and Culture

- Provide regular training on regulatory requirements and compliance best practices
- Foster a culture of compliance throughout the organization
- Implement systems for reporting and addressing potential compliance issues

Digital Quality in RDfR
Advanced Data Analytics

- Implement big data analytics to identify patterns and predict potential reliability issues
- Use machine learning algorithms to optimize design and manufacturing processes
- Develop real-time monitoring systems for product performance and quality

Digital Twin Technology

- Create digital twins of products to simulate performance under various conditions
- Use digital twins for predictive maintenance and reliability optimization
- Integrate digital twin data with product lifecycle management systems

Internet of Things (IoT) and Connected Products

- Implement IoT technologies to gather real-time data on product performance
- Use connected product data to inform future design improvements
- Develop predictive maintenance capabilities based on IoT data

Automated Quality Control

- Implement artificial intelligence (AI)-powered visual inspection systems
- Develop automated testing and validation processes
- Use robotics and automation to ensure consistent quality in manufacturing

Blockchain for Quality Assurance

- Implement blockchain technology for secure and transparent quality records
- Use smart contracts to automate quality control processes
- Ensure the integrity and traceability of quality-related data throughout the product lifecycle

4.3 Roadmap for Robust Design for Reliability Execution

To provide a roadmap and enhance the implementation of RDfR, we offer a practical RDfR roadmap procedure for guidance. This procedure serves as a reference to identify and define a set of reliability engineering activities necessary to meet the product's required reliability, functionality, lifetime, and total cost of ownership. It is formulated in broad terms, allowing its application across a diverse range of products in business-related applications. This document delineates the RDfR activities, checkpoints, and expected deliverables for projects.

This procedure is applicable to all stages of product development and covers RDfR activities aimed at ensuring reliability in the development of both hardware and software products for various applications. It is imperative that all teams involved in the product development lifecycle adhere to this procedure and its associated processes. Compliance with these guidelines is a shared responsibility among the teams to uphold reliability standards throughout the development process.

Figure 4.4 delineates the inputs, outputs, activities, decision points, and points for monitoring compliance within the RDfR procedure. Notably, specific details of the RDfR process are emphasized in bold. Additionally, definitions pertinent to this procedure are distinctly highlighted for clarity.

4.3.1 Identify Phase

The Identify phase encompasses both the identification and definition processes, including the Define stage. This phase holds significant importance in establishing the framework for reliability engineering throughout the product development lifecycle. It involves a thorough examination and clarification of customer requirements, acting as the cornerstone for subsequent reliability efforts.

The primary objective is to identify and articulate the fundamental elements of reliability, which include robustness and error prevention, as expressed in the equation [5]:

$$\text{Reliability} = \text{Robustness} + \text{Mistake Prevention} \tag{4.1}$$

This equation serves as a guiding principle, emphasizing the proactive nature of reliability engineering from the outset of the design process. Unlike conventional definitions that focus solely on the probability of product performance without failure within specified conditions and timeframes, our approach underscores sustained robustness and error prevention over time.

By defining reliability as the combination of robustness over time and error prevention, we not only acknowledge the importance of product performance but also prioritize the mitigation of potential failure points and errors throughout the product lifecycle. This proactive

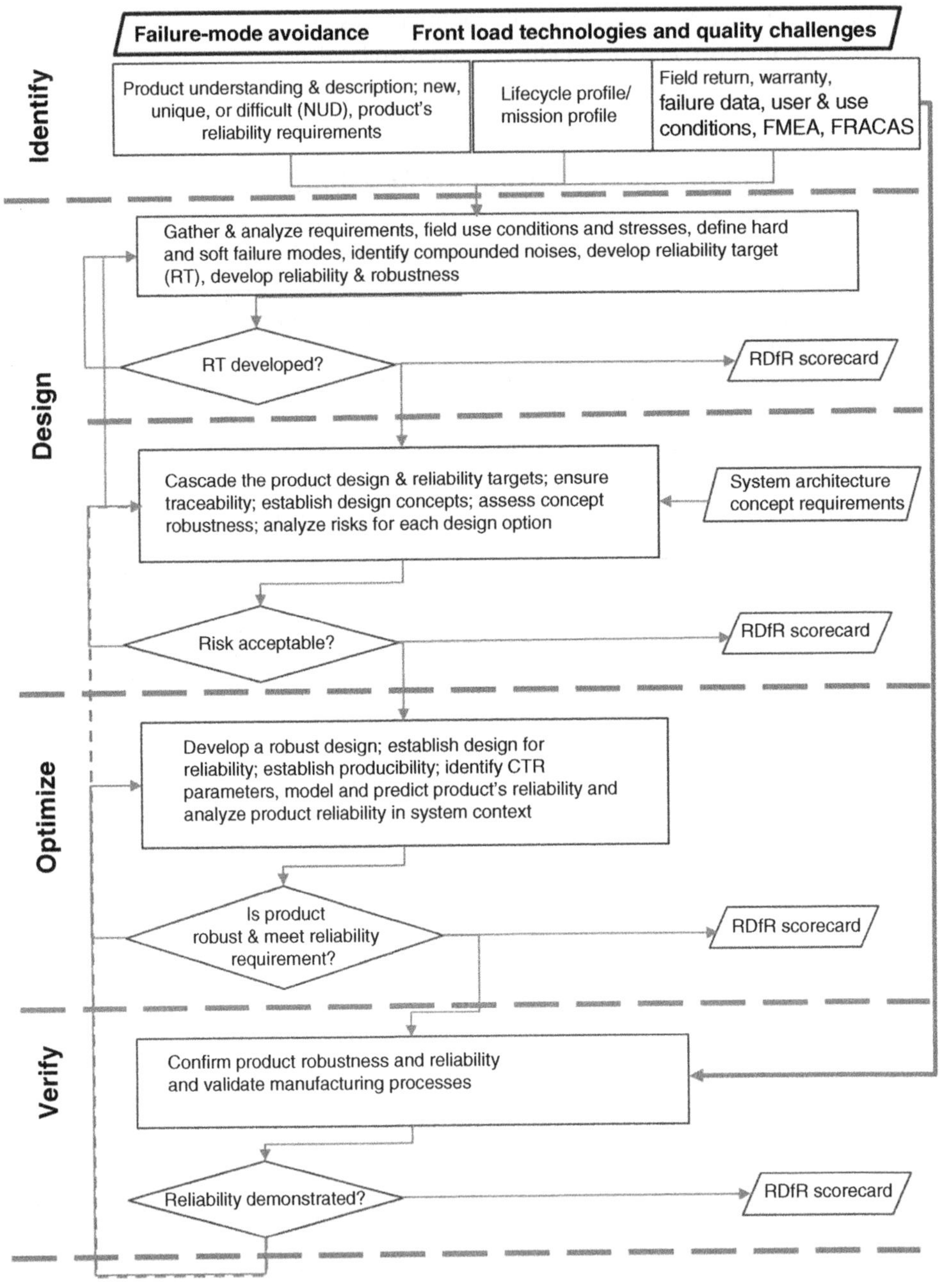

Figure 4.4 The RDfR Procedure Diagram.

stance encourages collaboration between product engineering and reliability engineering teams, fostering heightened confidence in the design and functionality of the product.

4.3.1.1 Identify Phase Purposes

A critical milestone in the Identify phase is establishing the product's reliability requirement. This requirement serves as a guiding beacon, directing subsequent engineering activities and ensuring the integration of reliability considerations into the fabric of the product's design from its inception. Through the incorporation of reliability requirements in the early stages of development, we establish a solid foundation for a product that is robust and dependable, meeting or exceeding customer expectations throughout its lifecycle.

To have a better understanding of the purpose of the chapter, let us review the following terminology and concepts:

Ideal Function: the primary purpose of the design is typically focused on energy-related aspects, as mechanical engineering revolves around either facilitating movement or impeding it.

Signal Factor: the energy input into the engineering system, whether by the customer or by an adjacent system, is what drives its operation or functionality.

Error State: an undesirable outcome of the engineering system, error states may manifest as failure modes. These can be categorized as either soft failures, involving variations in ideal function or degradation thereof, or hard failures, resulting in the complete loss of ideal function.

Control Factors: design parameters are aspects of the design that engineers have the ability to alter, such as dimensions, shapes, materials, positions, and locations.

Noise Factors: disturbances are origins of disruptive influences that can interfere with optimal functionality, resulting in error states that lead to quality issues.

- Robustness: refers to the reliable and consistent performance of ideal function around the target, remaining insensitive to the influence of noise.
- Parameter Design: adjusting control factors to discover improvements to Robustness.
- Failure Mode and Effect Analysis (FMEA) - manages the effects of failure modes,
- Failure modes are due to –
  - Mistakes
  - Sensitivity to noise factors.

The Role of Robust Engineering Design in Preventing Mistakes

Robust engineering design focuses on creating products and systems that perform reliably under a variety of conditions. However, while it significantly contributes to the overall reliability and durability of a product, it does not inherently prevent mistakes during the design and manufacturing processes. Mistake prevention requires a proactive approach that involves constant vigilance and systematic methodologies like Failure Mode and Effects Analysis (FMEA).

Utilizing FMEA for Mistake Prevention

Failure modes and effects analysis (FMEA) is a powerful tool for identifying and preventing potential failures. When integrated into the RDfR process, FMEA can significantly enhance an organization's ability to prevent mistakes and improve product reliability.

Systematic Evaluation of Failure Modes

- Conduct comprehensive FMEA for all critical components and systems
- Prioritize potential failure modes based on severity, occurrence, and detectability
- Develop and implement mitigation strategies for high-priority failure modes

Integration of Design Standards in FMEA

- Use design standards as detection events within the FMEA process
- Evaluate the effectiveness of current design standards in preventing failures
- Identify areas where design standards may need to be updated or enhanced

Continuous FMEA Process

- Implement FMEA as an ongoing process, not a one-time event
- Regularly review and update FMEA based on new data and insights
- Use FMEA results to drive continuous improvement in design and manufacturing processes

Preventing failure modes through vigilance is a critical aspect of RDfR. By understanding the entropic nature of mistakes, enhancing individual and organizational vigilance, utilizing tools like FMEA, and fostering a culture of continuous improvement, organizations can significantly enhance their ability to design and manufacture reliable products. This comprehensive approach to vigilance not only prevents failures but also drives innovation and efficiency, ultimately leading to superior product performance and customer satisfaction.

Integrating FMEA with Robust Engineering Design

While robust engineering design and FMEA serve different purposes as shown in Figure 4.5, they complement each other in enhancing the reliability and quality of products.

The P-diagram is a powerful tool in RDfR that combines functional and dysfunctional analyses into a single, comprehensive framework. This integration makes it particularly well suited for preparing design failure mode and effects analyses (DFMEAs) and can serve as an effective substitute for preliminary risk analysis across various engineering sectors, including mechanical, mechatronic, and electronic fields.

Integrating FMEA into the robust design process can significantly enhance mistake prevention efforts:

1) Proactive Risk Management: By incorporating FMEA early in the design process, potential risks can be identified and addressed before they become significant issues. This proactive approach ensures that the design is not only robust but also error-resistant.
2) Enhanced Design Reviews: Using FMEA as part of design reviews helps in systematically identifying and addressing potential failure modes. This enhances the thoroughness and effectiveness of design reviews.
3) Improved Communication and Collaboration: FMEA fosters better communication and collaboration among cross-functional teams. By involving various stakeholders in the risk assessment process, a more comprehensive understanding of potential risks is achieved.
4) Documentation and Knowledge Sharing: FMEA provides detailed documentation of potential risks and mitigation strategies. This serves as a valuable knowledge base for future projects and helps in avoiding past mistakes.

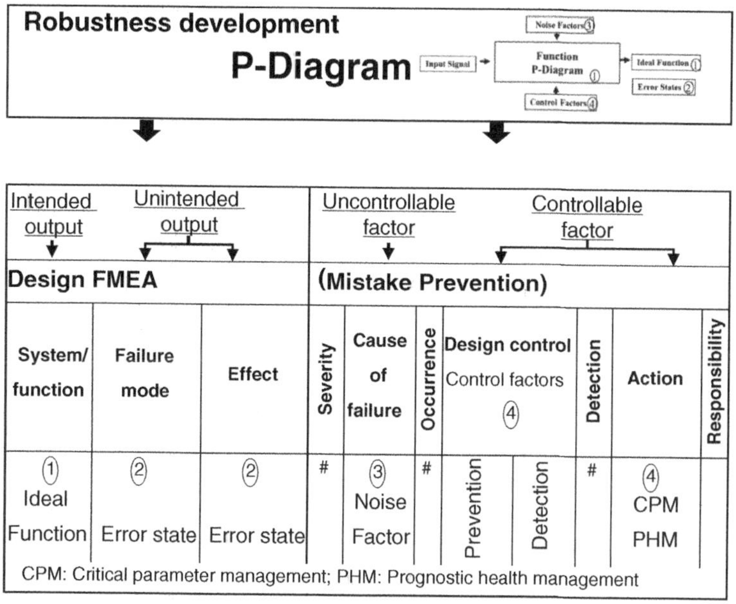

Figure 4.5 P-Diagram and DFMEA Mapping.

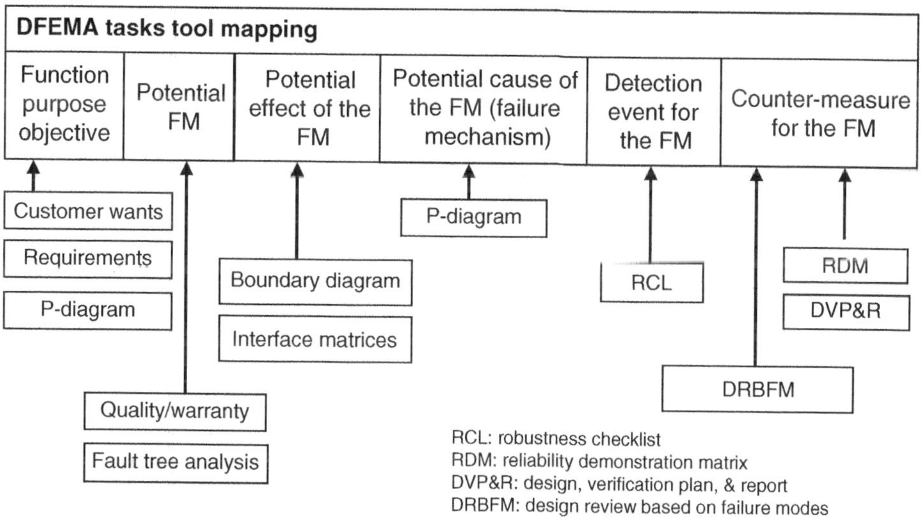

Figure 4.6 DFMEA Tasks and Tool Mapping.

Robust engineering design improves product reliability and performance by reducing sensitivity to variation, but it does not by itself eliminate design mistakes or failure risks. As shown in Figure 4.6, mistake prevention requires a structured DFMEA process supported by specific analytical tools at each stage. The process begins with clarifying the function purpose/objective, which is derived from customer wants, requirements, and the P-diagram. From there, potential failure modes are identified, often using sources such as quality/warranty data and fault tree analysis. The potential effects of failure modes are then better understood through tools like the boundary diagram and interface matrices, which help reveal how subsystem interactions and external interfaces can generate unintended effects.

To determine the potential causes of failure modes, the P-diagram plays a central role by linking functional intent with noise factors, error states, and failure mechanisms. Once these causes are understood, the detection event for the failure mode can be established using tools such as the RCL (robustness checklist), while countermeasures are strengthened through methods including DRBFM, RDM, and DVP&R, which support design reviews, reliability demonstration, and verification planning. In this way, Figure 4.6 shows that DFMEA is not an isolated activity, but a coordinated framework that connects robust design tools with risk identification, failure analysis, detection planning, and corrective action.

Therefore, integrating robust engineering design with DFMEA and its associated tool set creates a more complete approach to mistake prevention. Robust design improves consistency and performance, while DFMEA ensures that possible failure modes, their causes, their effects, and their controls are systematically identified and managed. This integration leads to designs that are not only reliable and high performing, but also more resistant to errors, easier to validate, and better aligned with customer expectations. Ultimately, such a combined approach supports higher product quality, lower warranty and development costs, improved customer satisfaction, and stronger competitive advantage.

4.3.1.2 Identify Phase Activities

Reliability by robust design to avoid failure modes and mistakes prevention. The RDfR activities outlined in the Identify phase serve as a reference point for deliberation.

1) Front load new technologies and challenges
2) Gathering requirements and classifying for difficulty to develop
   When requirements have been identified, classify into two buckets [6]:

- New, Unique, Difficult (NUD) requirements*
  - New: a requirement that has never been done by you or your competitors.
  - Unique: a requirement like ones that exist but has unique aspects this time that are challenging.
  - Difficult: a requirement that is very hard to fulfill and that requires special focus to deliver well.
- Easy, Common, Old (ECO) requirements*
  - Easy: a requirement is not difficult for you to fulfill and has low risk of failing to be done.
  - Common: a requirement that you and competitors have done before with a repeatable history of success.
  - Old: a requirement you have a great deal of experience fulfilling but others may not.

3) Input and analyze the product's requirements.

---

* NUD and ECO are concepts coined by Skip Creveling.

4) Identify noise factors or stressors for a design of failure function, input and analyze the product's actual use conditions.
5) Input and analyze return of experience, i.e., field reliability data and failure analyses from relevant legacy products and candidate components.
6) Identify and choose noise factors as presented in Table 4.2 and correlate these factors with the bathtub curve as illustrated in Figure 4.7.

Table 4.1 Example of Noise Factors for Failure Function.

Consider	Examples	Caused by
Customer	• Usage Rate/Plot density & Duty Cycle/Idle times	Conditions of use
Operation	• External Operating Environment/Energy/Tilt	Climatic and application conditions
Integration	• Ink variations/Servicing stresses/Other design variations under consideration/Error states from another component	Component and system interactions
Mfg. & Supply Chain	• Process variations/Material properties, contamination/Defects we would like designs to be robust to	Low-cost mfg, production rate
Wear-out	• Changes over time in dimensions or strength (such as wear out, fatigue, deterioration, chemical degradation)	Exposure to repetitive demand

Table 4.2 Types of Noise Factor that Disrupt Ideal Function.

	Types of Noise Factor that Disrupt Ideal Function:
Inner Noises	1. Piece-to-piece variation of part properties (such as dimensions) due to production conditions
	2. Changes over time/mileage in dimensions or strength (such as wear out, fatigue, chemical degradation)
Outer Noises	Conditions of use
	3. Customer usage and duty cycle
	Operating environment
	4. External (climatic and road conditions)
	5. Internal, due to system interactions, i.e., error states being received as noise factors from neighboring sub-systems

Note: the effects of noise factors can often be represented by others, via transfer functions—for example, forces can be converted to geometry. This can lead to more efficient testing.

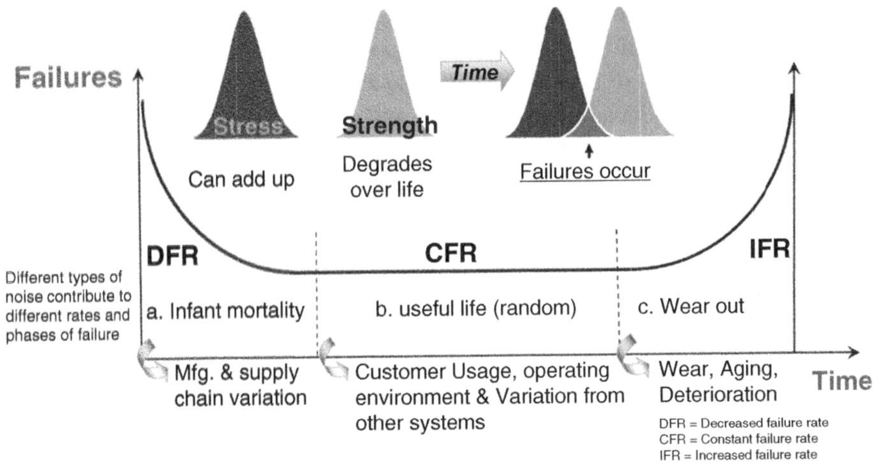

Figure 4.7 Bathtub and Noise Factors Mapping.

Noise factors can be leveraged to improve reliability with higher engineering confidence through robust design strategies. Instead of attempting to eliminate noise factors, engineers can design systems that are less sensitive to these variations and disturbances. By understanding the impact of noise factors on the performance of a product or system, engineers can develop design solutions that mitigate the effects of variability. This involves identifying critical parameters, incorporating redundancy, implementing fault-tolerant mechanisms, and optimizing designs to enhance overall robustness.

Robustness Metrics

All robustness metrics have three things in common.

- They can always be presented graphically.
- They always respond to noises.
- They measure the distance from the failure modes, in units of physics, geometry, and properties of materials.

The robustness metrics have a natural hierarchy - from least useful to most useful; these are

- Bogey testing metrics (e.g., % passed), which is not recommended based on the three criteria for a robustness metric given above.
- Time-to-failure metrics (e.g., Weibull plots)
- Degradation metrics
- Metrics based on a measure of the Signal relative to a measure of the Noise** (from the P-diagram)

Through RDfR, engineers proactively address potential challenges posed by noise factors during the early stages of product development. This approach allows for the creation of products that can consistently perform well under various conditions, providing a higher level of confidence in the reliability of the final product. The focus is on building resilience, adaptability, and stability into the design to ensure that the system can withstand variations and uncertainties in real-world environments.

7) Input and analyze system-level functional, architectural, and interface requirements.
8) Establish the product's reliability requirement target using the above definition of reliability together with appropriate quality and reliability techniques.
9) Perform benchmarking of the new requirements by comparing them to those of existing products including, wherever possible, a comparison with the competition. Document the results of this change point analysis in a summary of the comparisons that highlights risks.
10) Consider following discovery level stress guidelines as shown in Figure 4.8:

- Avoid excess overstress.
- Use to develop and establish margin.
- Enables short, focused tests—speeds learning cycles.
- Do not need to pass discovery level stress tests.
- If designs pass, less characterization is needed.

11) Develop initial P-diagram. See details for P-diagram development in design phase.
12) Start a Design Failure Mode and Effects Analysis (DFMEA) for the design concept(s). This functional DFMEA shall identify potential failure modes, their probability of occurrence and severity of their consequences, from the customer's point of view, along with mitigation and contingencies.
13) Monitor, on a regular basis using the RDfR scorecard, the progress made by the project team to complete key RDfR tasks and to integrate the tasks into their project.

Identify RDfR Decision Point: A key decision must be made on whether the risks are acceptable. This decision should be based on the completed RDfR deliverables, justifying that the risks have been adequately identified and addressed by the engineering plan gateway review.

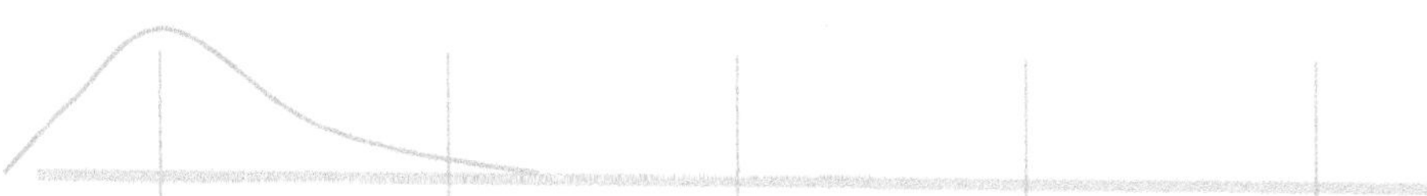

Nominal	Characterization	Just Beyond Specs	Discovery level	Excess Overstress
• Represented by expected distribution of stressor levels • No special effort to build or test stressor extremes • Requires large sample sizes to validate performance	• Near the bounds of expecting operating conditions • Covers spec limits • Use to validate capability at spec limits	• Just outside of spec requirements • Use when difficult to build & test other stressor vectors	• Highly accelerated stresses but avoid levels that shift to non-representative failure modes • Use to find and fix defects • Use to select most robust designs	• Characterized by shift to failure modes not expected in normal operating range • Lose ability to differentiate design robustness – everything fails

Figure 4.8 Test Stress Level Guidelines

4.3.1.3 Identify Phase Deliverables

Deliverable	Monitoring
Reliability Requirement	RDfR
Clear statement of use conditions	RDfR
Clear statement of lifecycle profile	RDfR
Quality history	RDfR
Initial P-diagram	RDfR
DFMEA for the most critical risks	RDfR
Front load new technologies and challenges	RDfR
New, Unique, Difficult (NUD) requirements	RDfR
Easy, Common, Old (ECO) requirements	RDfR
PHM	

4.3.2 Design Phase

The design phase in RDfR is a critical step aimed at developing products that consistently perform well under a variety of conditions and withstand the test of time. Robust design focuses on creating solutions that are not only functional but also resilient to variations in the manufacturing process and environmental factors. The Design phase involves several key activities:

1) Defining Reliability Requirements: Understanding and translating customer needs and reliability standards into specific, measurable design criteria.
2) Concept Generation and Evaluation: Developing and assessing multiple design concepts to identify the most robust and reliable options.
3) Detailed Design and Analysis: Creating comprehensive design specifications, including materials, components, and assembly processes, with a focus on reliability and robustness.
4) Failure Modes and Effects Analysis (FMEA): Systematically identifying potential failure modes and their effects and implementing strategies to mitigate these risks.
5) Tolerance Design: Determining acceptable levels of variability in product dimensions and performance attributes to ensure consistent quality.
6) Design Verification and Validation: Testing prototypes and models under various conditions to ensure the design meets all reliability requirements and performs consistently.

Throughout the design phase, specific tools and techniques such as DOE, parameter design, tolerance design, and control charts are employed to optimize the design for reliability [7, 8]. These methods help in systematically identifying and controlling factors that affect product performance, ensuring that the final design is both robust and reliable.

By rigorously following these steps, organizations can develop products that deliver consistent performance and high reliability, even in the face of variability and unforeseen challenges. This leads to enhanced customer satisfaction, reduced warranty costs, and a stronger competitive position in the market.

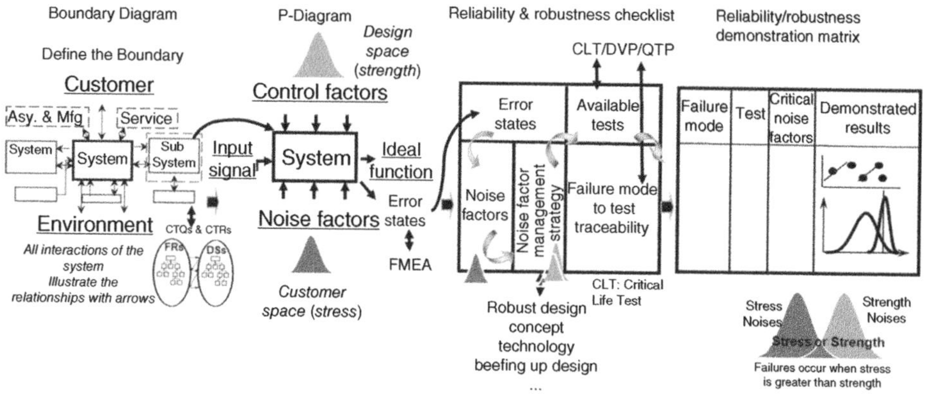

Figure 4.9 Flow of the Discipline Tasks.

4.3.3 Design Phase Purposes

The purpose of the design phase in RDfR (IDOV) is to develop products that maintain high performance and dependability under a wide range of conditions and throughout their intended lifespan. This phase is crucial for ensuring that the final product is not only functional but also resilient to variations in manufacturing and environmental factors. Complete an engineering quality discipline, as shown in Figure 4.9, that minimizes the risks in developing and producing a robust product whose predicted reliability meets the requirements.

Key objectives of this phase include:

1) Defining Reliability Objectives: Clearly establishing the reliability targets and performance criteria based on customer needs and industry standards.
2) Developing Robust Concepts: Creating and evaluating design concepts that can withstand variability and minimize the impact of uncontrollable factors.
3) Detailed Design Specifications: Translating the selected concept into detailed engineering designs, including materials, dimensions, and processes that ensure reliability.
4) Identifying and Mitigating Risks: Using tools such as FMEA to proactively identify potential failure points and implement strategies to prevent them.
5) Ensuring Consistency and Quality: Applying techniques like tolerance design and DoE to ensure the product performs consistently and meets quality standards despite variability.
6) Verification and Validation: Conducting rigorous testing and analysis to confirm that the design meets all reliability requirements and performs reliably under expected operating conditions.

By focusing on these objectives, the design phase aims to create products that are robust against variability and demonstrate high reliability in real-world use. This not only enhances customer satisfaction but also reduces the likelihood of costly failures and warranty claims, thereby improving overall product success and longevity.

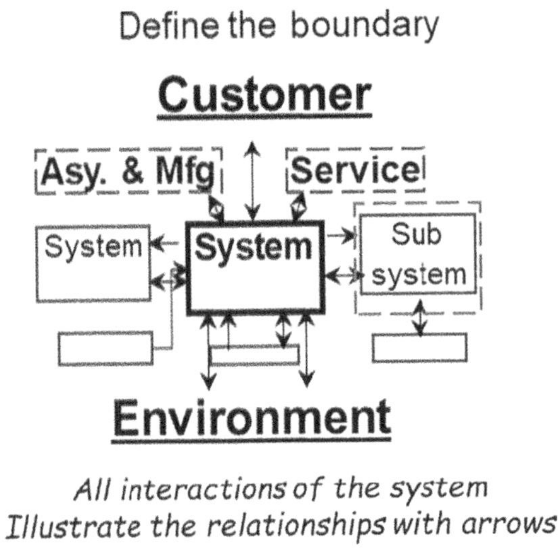

Figure 4.10 Boundary Diagram.

Develop boundary diagram as shown in Figure 4.10 to scope the boundary for reliability effort. The purpose of the boundary diagram is to serve as a:

- Graphical tool for defining the system's scope.
- Means of capturing internal and external system interfaces.
- Specification of ownership and the cascade of requirements to interfacing systems.

As the saying goes, "A picture is worth a thousand words."

Develop or Update Robustness Model P-Diagram The function P-diagram serves as a tool for engineering teams to gain a better understanding of the energies impacting their systems. It helps them identify and mitigate undesirable outputs that could result in degradation or the loss of ideal function. This tool represents an initial phase in the creation of a reliability and robustness checklist (RRCL). The noise factors and error states pinpointed in the P-diagram will undergo further analysis in the RRCL.

The function P-diagram effectively captures the team's experience, highlighting potential areas of concern relevant to their systems. It aids in the development of robustness tests and parametric design, with a focus on describing parameters that dictate how a design interacts with its environment, both internally and externally.

When the product is used in the customer's environment, stress is induced to the system. When that stress of the product environment is greater than the strength of the design, we may get error states out of the system instead of the ideal function. If we can analyze our design and predict these error states, we can strengthen our design and produce a more robust product for our customers. The stress–strength Interference principle states that a product fails when the stress experienced by the product exceeds its strength (as shown in Figure 4.11).

In order to reduce the failure probability (and thus increase reliability), we must reduce the interference between stress and strength. A structured process, such as the RDfR process presented in this chapter, is needed to achieve this.

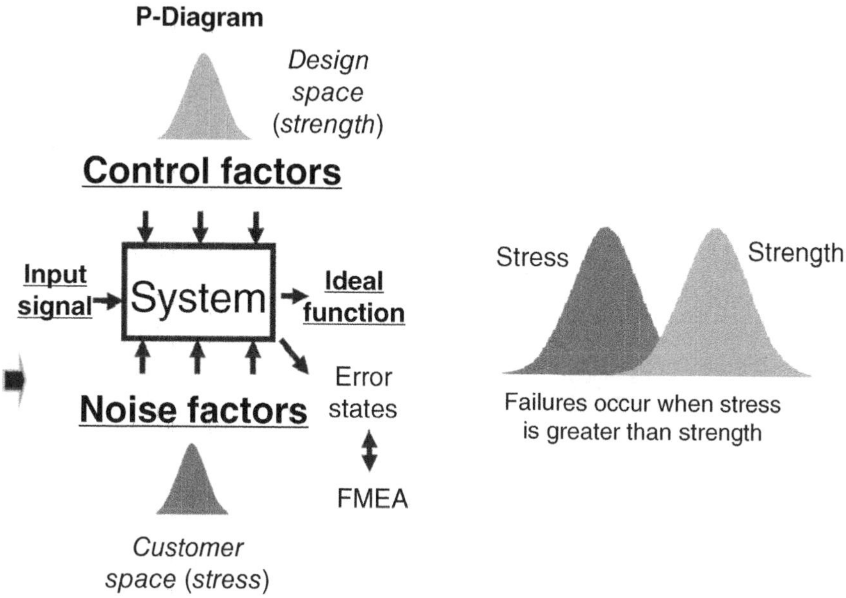

Figure 4.11 P-Diagram and Strength and Stress Analysis.

The control factors embody the robustness of the design, while noise factors represent the potential stressors.

It's crucial to optimize the design so that its strength significantly surpasses the stress levels; otherwise, failure modes and errors may arise due to the influence of these stressors. The development of resilient products hinges on effectively analyzing and designing the system. A hierarchical set of requirements delineates the functions at lower levels that necessitate optimization to fulfill higher-level system requirements. Employ creativity to deliver system solutions for each function that are of high quality, reliability, and robustness. Detailed guidance on constructing a meaningful P-diagram is already provided in Chapter 2.

1) Update the DFMEA with any new failure modes discovered during this phase and take appropriate action in the design to minimize their occurrence and mitigate their severity.
2) Assess concept robustness.
3) Analyze the risks for each design concept.
4) Develop RRCL as shown in Figure 4.12 to identify and prioritize key functions and high-impact error states as potential reliability demonstration plan and report (RRDM) candidates for further analysis in RRDM.

Purpose of RRCL

- Improve team communication (a planning tool to capture engineering knowledge).
- Identify all relevant noise factors.
- Establish noise factor–error state relationships and linkage to FMEA.
- Establish relationship between tests (design verification methods) and the identified error states and noise factors.

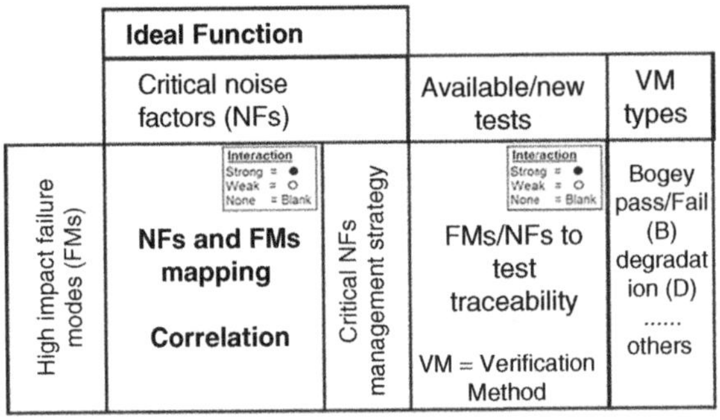

Figure 4.12 Example of Reliability and Robustness Checklist (RRCL).

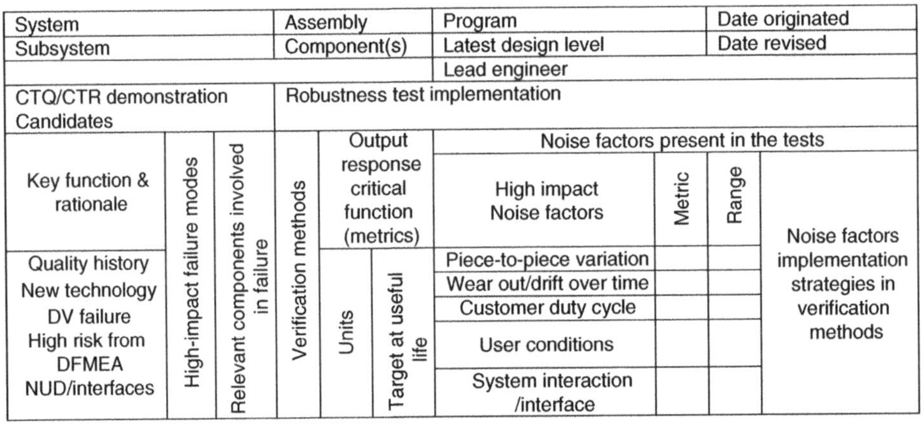

Figure 4.13 Reliability and Robustness Demonstration Plan Heading.

- Identify the need for new tests, helps develop new tests.
- Identify redundant tests.

The main outputs of RRCL are:

- High-impact error states for further analysis in reliability demonstration matrix (RDM) as shown in Figure 4.13.
- High-impact noise factors for the demonstration tests in RRDM.
- High-impact test verification methods for the robustness demonstration tests in RRDM.

Conduct robust parameter design or experiments to explore, estimate, or confirm the reliability of critical components, subassemblies, or complete prototypes as appropriate. Use techniques such as modeling and simulation, DOE to identify noise management strategy, determine the sensitivity, interactions, and optimal values for design parameters in the presence of uncertainty. To discover any unknown risks and failure modes, provoke failure by continuing testing beyond the required limits. Conduct root cause analysis (RCA) or equivalent of those failures with assistance from appropriate technology experts and their laboratory facilities.

Develop design verification plans/methods (DVP/DVM). Facilitate the creation of a thorough design verification plan (DVP) that validates the product's reliability requirements and tackles each significant risk identified by the RRCL and the revised DFMEA.

Figure 4.14 Reliability and Robustness Demonstration Matrix.

Reliability and robustness demonstration matrix

Failure mode	Test	Critical noise factors	Demonstrated results
			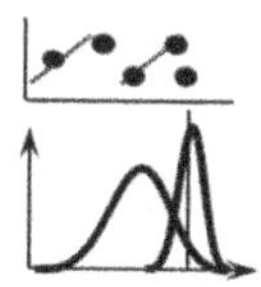

Purpose: Ensure the update and improvement of all tests and test plans as needed.

Develop reliability and robustness demonstration matrix (RRDM) as shown in Figure 4.14.

Purpose: Demonstrate that the system / component performs its intended function over its useful life in the presence of noise factors, i.e., is robust and reliable.

Analyze potential production processes and identify significant variability or sensitivity issues that may adversely affect the product's manufacturing yield, reliability, or maintainability.

Recommend appropriate PHM based on the CTRs parameters findings from the robust optimization the DFMEA, design analyses, and test results.

Recommend an appropriate field reporting and corrective action system (FRACAS) for the product based on its DFMEA and PHM together with the return of field experience.

Monitor, on a quarterly basis using the RDFR scorecard, the progress made by the project team to complete key RDFR tasks and integrate them into their project.

4.3.3.1 Design Phase Deliverables

Deliverable	Responsibility	Monitoring
Reliability Requirement		RDfR
Clear statement of use conditions		RDfR
Clear statement of lifecycle profile		RDfR
Summary of field reliability experience for similar products and components		RDfR
P-diagram for critical output response(s)		RDfR
DFMEA for the most critical risks		RDfR
Boundary diagram		RDfR
Noise Factor Management Strategy		RDfR
Reliability and Robustness Checklist		RDfR
Reliability and Robustness Demonstration Matrix		RDfR
Root Cause Analysis (RCA)/8D/DMAIC or equivalent		RDfR
PHM recommendations		RDfR

4.3.4 Optimize Phase

The optimize phase in RDfR focuses on refining and enhancing the product design to ensure maximum robustness and reliable performance. This phase is critical for identifying and implementing improvements that mitigate risks, reduce variability, and ensure the product performs consistently under diverse conditions. The key objectives of the Optimize phase include:

1) Fine-tuning Design Parameters: Adjusting design parameters to achieve optimal performance and reliability, ensuring the product meets all specifications and customer expectations.
2) Reducing Variability: Identifying sources of variability in the manufacturing process and product operation and implementing strategies to minimize their impact.
3) Enhancing Robustness: Applying advanced design techniques to make the product more resilient to environmental and operational variations.
4) Risk Mitigation: Using tools such as FMEA to proactively address potential failure points and ensure that the design changes enhance reliability.
5) Verification and Validation: Conducting thorough testing and analysis to confirm that the optimized design meets all reliability requirements and performs consistently under real-world conditions.

During the optimize phase, various tools and methods, such as DOE, Taguchi methods, and response surface methodology (RSM), are employed to systematically evaluate and improve the design [9]. By applying these techniques, the Optimize phase aims to refine the product to its highest level of robustness, reliability, and performance, ensuring long-term success and customer satisfaction.

4.3.4.1 Optimize Phase Purpose

In the optimize phase of RDfR, the objectives aim to achieve a robust and reliable product design through:

- Minimizing sensitivity to variations in materials, components, and processes.
- Discovering the design margins to failure and root causes of significant failures.
- Verifying by testing and analysis that the product's required reliability is achieved.
- Predicting the product's manufacturing yield [10].

Within the RDfR discipline toolset workflow, the failure-mode avoidance and robustness review of a design involves the use of the P-diagram, RRCL, and reliability and robustness demonstration plan.

As we discussed in Chapter 2, the following P-diagram paradigm is used as the basic model for Taguchi method [8]:

Figure 4.15 illustrates a general design model, where $y = (y_1, \ldots, y_u)$ is the output vector that stands for the system performance, or the left hand of the transfer functions vector, $x = (x_1, \ldots, x_n)$ is the design parameter (or the process variable) vector of the design; $z = (z_1, \ldots, z_p)$ is the vector that stands for the uncontrollable factors, or noise factors of the design.

Clearly, both design parameters and noise factors play a role in influencing the output y of the transfer function. In this chapter, our focus is solely on a specific requirement

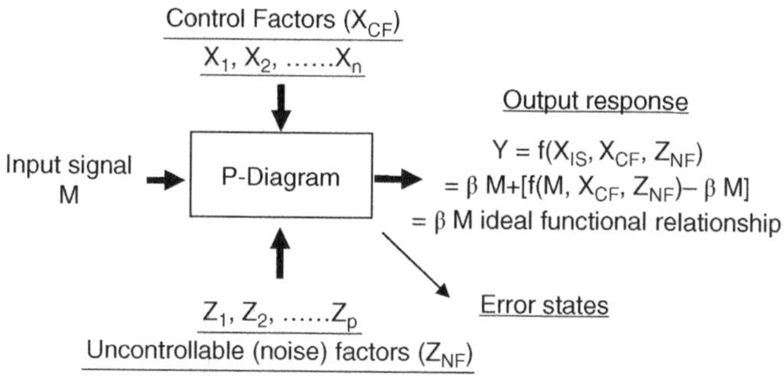

Figure 4.15 P-diagram.

characteristic, denoted as y for the time being. In real-world scenarios, it is common to address multiple requirements simultaneously. The subsequent transfer function is utilized to depict the cause-and-effect relationship between the output y and the variables x and z.

$$y = g(x_1,\ldots,x_n, z_1,\ldots,z_p) \tag{4.1}$$

In some circumstances, we can assume that output y is determined by design parameters x, and the design parameters are influenced by noise factors, z. So x = d(z), in this case:

$$y = f(x_1,\ldots,x_n) \tag{4.2}$$

Mathematically, $z = (z_1,\ldots,z_p)$ is a random vector with unknown distribution. The design parameters $x = (x_1,\ldots,x_n)$ is also a random vector, because in mass production we cannot guarantee to produce massive number of identical products, there will be piece-to-piece variation. We can assume x_i, for i = 1 to n. the nominal value of design parameter can be chosen by the designer, we can define, which is the standard deviation of design parameter, it can also be controlled (with a cost) at tolerance design stage.

Because both x and z are random variables, the requirements, Y, will of course be a random variable. In Taguchi's parameter design, it is assumed that there is a target value for Y, T. For example, the main basic function of a power supply circuit is to provide electrical power to small appliances, a key output performance is its output voltage. If the needed voltage for smaller appliance is 16 volt, then the target value T = 16.

Parameter design is the stage after the system design. In the parameter design stage, the transfer function has already been finalized in the system design stage, but the nominal values of design parameters can still be adjusted. For example, when we design power supply circuits for small appliances. After system design, the circuit diagram has already been finalized. But we still can change the design values of the circuit parameters, such as the designed value of resistors and transistors.

Given the transfer function:

$$Y = g(x_1, x_2, \ldots, x_n; z_1, z_2, \ldots, z_n). \tag{4.3}$$

We can get the following relationships by Taylor series approximation:

$$\Delta y = \frac{\partial g}{\partial x_1}\Delta x_1 + \frac{\partial g}{\partial x_2}\Delta x_2 + \cdots + \frac{\partial g}{\partial x_n}\Delta x_n + \frac{\partial g}{\partial z_1}\Delta z_1 + \frac{\partial g}{\partial z_2}\Delta z_2 + \cdots + \frac{\partial g}{\partial z_m}\Delta z_m \tag{4.4}$$

and

$$Var(Y) = \sigma_y^2 \approx \left(\frac{\partial g}{\partial x_1}\right)^2 \sigma_{x_1}^2 + \left(\frac{\partial g}{\partial x_2}\right)^2 \sigma_{x_2}^2 + \cdots + \left(\frac{\partial g}{\partial x_n}\right)^2 \sigma_{x_n}^2 + \left(\frac{\partial g}{\partial z_1}\right)^2 \sigma_{z_1}^2 + \left(\frac{\partial g}{\partial z_2}\right)^2 \sigma_{z_2}^2 + \cdots \left(\frac{\partial g}{\partial z_m}\right)^2 \sigma_{z_m}^2 \tag{4.5}$$

From Eq. (4.3–4.5), it is clear that we can reduce Var(Y) either by reducing the sensitivities $\frac{\partial g}{\partial x_i}$, for i = 1 to n, which are the sensitivities to the variation in design parameters, or $\frac{\partial g}{\partial z_j}$, for j=1 to m, which are sensitivities to noise factors.

Fortunately, many of these sensitivities are influenced by the nominal values of design parameters. Figure 4.16 shows how parameter setting may influence the variation of Y [11]. Both transfer function y = g(x,z) and sensitivity, for example, $\frac{\partial g}{\partial x_i}$, can be nonlinear functions of parameter setting, therefore, by choosing the parameter setting at low sensitivity area, we can reduce Var(Y) greatly though we have the same variation of either design parameters or noise factors.

Changing the nominal of x will have a major effect on the variability of y.

Equations (4.3–4.5) indicates two distinct areas of focus for improving designs as follows:

- Minimize input sigma σ_i (variability), i from 1 to n by improving process capability. This is the kind of Six Sigma DMAIC project.
- Minimize sensitivities $\frac{\partial g}{\partial x_i}$ (sensitivity), i from 1 to n by selecting optimal nominal levels through robust design. The efforts of reducing sensitivities will be much more effective in early design phase. That is why most of the proactive robust design projects are in the category of RDFR projects. See Figure 4.17. This is very important and that is why we use the partial derivatives of the xs to define the Ys. Of course, if the transformation function is a linear one then the only thing we can do is to control variability.

It goes without saying that in the majority of cases, we encounter polynomials, underscoring the significance of DOE and particularly parameter design in any RDfR initiative.

In summary, the tasks of parameter design are as follows:

a) Adjusting the nominal settings of design parameters to reduce the sensitivities of functional performance y to the variations of design parameters and noise factors, as a result, reducing the system functional performance variation Var(Y).
b) Adjusting the nominal settings of design parameters to shift mean performance level to its target value T.

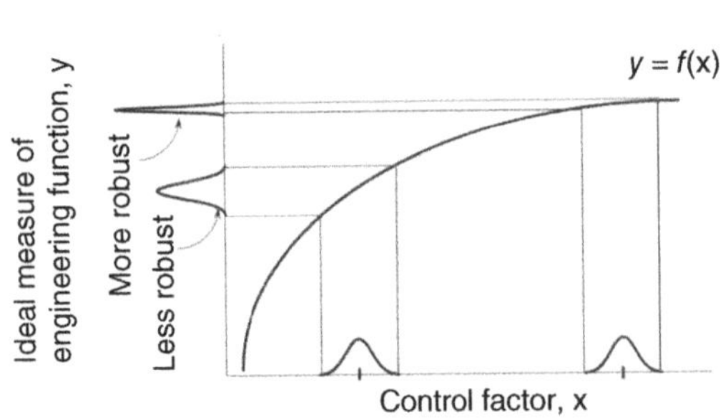

Figure 4.16 Adjusting Sensitivity to Reduce Var(Y).

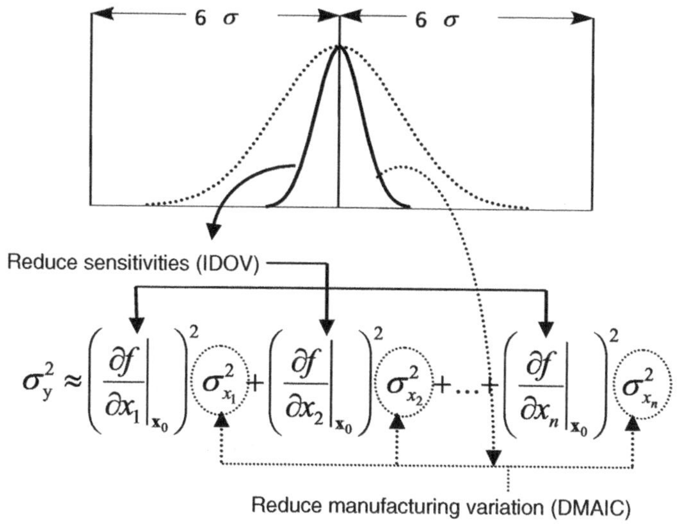

Figure 4.17 Total Variation Reduction.

The major objective of task (a) is to make the system insensitive to all sources of variations; it is also called to make system robust to all sources of variations. Therefore, Taguchi's parameter design is also called robust parameter design.

4.3.4.2 Robustness "Rules of Engagement"

1) Concentrate on Ideal Function and establish a way to measure it; do not use symptoms of poor quality to measure robustness.
2) Identify the sources of the five types of noises and expected magnitudes (remember noise 5).
3) Concentrate on the effects of the noises; maybe one noise can be used to represent others.
4) Understand how error states and noise factors cross system interfaces and boundaries. Establish contracts with neighboring teams.
5) Develop a noise factor management strategy. Removing the noise might be easier than becoming robust to it. The laws of physics are strict.
6) Work out how to include remaining noise factors in all tests in the DVP.
7) Plan a robustness assessment of current design to compare against ideal performance.
8) Where robustness improvement strategy is obvious from knowledge of physics, DO IT!
9) Where robustness improvement is not obvious, plan parameter design studies (using DoE if necessary) to discover the improvement.

The P-diagram serves as an early stage in creating a robustness checklist. Identified noise factors and error states in the P-diagram undergo further analysis in the robustness checklist, capturing the team's experience to pinpoint potential concerns relevant to their systems.

Following the P-diagram creation, the information becomes inputs for conducting DFMEA. The RRCL documents how the system functions with inputs and outputs.

The RRCL ensures proactive communication, planning, and budgeting in the upfront design phase, preventing reactive responses to problems post-product launch. Systematically using the checklist to devise a robust test plan identifies candidates for further analysis in the reliability and robustness demonstration plan and report.

The goal is to significantly reduce unnecessary redesign cycles, enabling a successful launch and focus on a second-generation design. This approach also cuts down product development time while increasing engineering confidence.

Lastly, the reliability and robustness demonstration plan captures and displays key verification methods, including high-impact noise factors, necessary to verify system robustness and reliability. It provides the team with a means to exhibit and demonstrate objective assurance of system robustness.

You may also be using the checklist in parallel with DFMEA activities leading up to the DVP.

The checklist analysis also supports your definition of the control plans for critical functional responses and critical design parameters in the Sustainability phase.

4.3.4.3 Optimize Phase Activities

Optimize Phase Activities in Robust Design for Reliability

The optimize phase in RDfR focuses on refining and enhancing product design to achieve optimal reliability and performance by reducing sensitivity to variations. This phase involves several critical activities:

1) **Identify Sensitivities:** Identify items highly sensitive to variability in materials, environment, manual techniques, or manufacturing processes.
2) **Perform Sensitivity Analysis:** Conduct sensitivity analyses to establish upper and lower control limits (UCLs and LCLs) for these parameters or processes.
3) **Consider Alternative Designs:** If an acceptable level of variability cannot be achieved, explore alternative design choices.
4) **Document Lessons Learned:** Capture and share significant lessons learned by documenting them in clear and concise technical memos and 8D or problem cause solution equivalents. Ensure these documents are searchable and retrievable by other project teams.
5) **Monitor Progress:** Regularly use the RDfR scorecard to monitor the project team's progress in completing key RDfR tasks and integrating them into the project.

By systematically carrying out these activities, the optimize phase ensures that the final product design is not only robust and reliable but also producible at scale with consistent quality and efficiency. This leads to a high-quality product that meets customer expectations, reduces warranty costs, and enhances overall competitiveness in the market.

Key activities in the optimize phase include:

1) Parameter Optimization
2) Mistake Prevention and Risk Mitigation
3) Variability Reduction
4) Robustness Enhancement
5) Verification and Validation

6) Design for Producibility
7) Cost Optimization

1) Parameter Optimization:
   - Taguchi Methods: Applying robust design techniques to find the optimal settings for design parameters that make the product less sensitive to variations.
   - Design of Experiments (DoE): Conducting structured experiments to identify the best combination of design parameters that maximize performance and minimize variability.
2) Mistake Prevention and Risk Mitigation:
   - Failure Modes and Effects Analysis (FMEA): Continuously updating FMEA to identify new potential failure modes and implement strategies to mitigate these risks.
   - Fault Tree Analysis (FTA): Analyzing the root causes of potential failures and developing preventive measures.
3) Variability Reduction:
   - Process Capability Analysis and development: Assessing and improving the capability of manufacturing processes to produce parts within specified tolerances.
   - Statistical Process Control (SPC): Implementing SPC tools to monitor and control the manufacturing process, ensuring consistent product quality.
4) Robustness Enhancement:
   - Tolerance Design: Determining the optimal tolerance levels for various components to ensure reliable performance without excessive cost.
   - Environmental Testing: Conducting tests under different environmental conditions (e.g., temperature, humidity) to ensure the product maintains reliability in diverse scenarios.
5) Verification and Validation:
   - Prototype Testing: Building and testing prototypes to validate that the optimized design meets all performance and reliability requirements.
   - Reliability Testing: Performing accelerated life testing and other reliability tests to predict the product's lifespan and identify any potential issues.
6) Design for Producibility:
   - Design for Manufacturability (DFM): Ensuring the design is optimized for efficient and cost-effective manufacturing.
   - Lean Manufacturing Techniques: Applying lean principles to streamline production processes and eliminate waste.

   The Optimize phase ensures that the product design is not only robust and reliable but also producible at scale with high quality and efficiency. This integrated approach leads to a product that meets customer expectations for reliability and performance while being cost-effective and feasible to manufacture, ultimately enhancing competitiveness and customer satisfaction.
7) Cost Optimization:
   - Cost–Benefit Analysis: Balancing the cost of design improvements with the benefits of increased reliability to ensure an economically viable product.
   - Value Engineering: Analyzing product functions to reduce costs without compromising quality and reliability.

4.3.4.4 Optimize Phase Deliverables

Deliverable	Responsibility	Monitoring
Updated Parameter Studies		DfR scorecard
Updated A list of CTQs and CTRs		
Updated Reliability Statement of Work (RSOW)		
Updated DFMEAs		
Updated Reliability and Robustness Checklist		
Updated Robustness Plan		
Updated Noise Factor Management Strategy		
Update PFMEA with manufacturing experts		RDfR scorecard
Summary of RRDM results		RDfR scorecard
Summary of accelerated testing, design margins and root cause analysis (RCA) of significant failures		
Implemented PHM as applicable		
Critical Parameters Management (CPM) plan		
Design Review based on Failure Modes (DRBFM)		
FRACAS recommendations		

Activities/tasks accomplished leading up to the Verification phase are:

- Design optimized using analytical tools or hardware DoEs.
- Optimized designs confirmed analytically or using hardware.
- Significant control factors and their levels known.
- Significant noise factors known.
- Significant noise X control interactions known.
- The verification phase is where we confirm that the product not only performs as required by the customer without creating any new customer dissatisfaction issues.

What is unknown so far?

- Potential Failure Modes Due to Interfaces, New Changes.
- Durability of the design.
- Interactions with other system attributes—e.g., optimized for NVH, what does that mean to fuel economy, durability, and performance?
- Effect of system interactions not included in the optimization study.

4.3.5 Verify Phase

4.3.5.1 Verify Phase Purpose in Robust Design for Reliability

The Verify phase is crucial for ensuring that the product design meets all reliability and performance requirements before full-scale production begins. This phase ensures the final design aligns with the PDP, meets customer expectations, and adheres to quality and reliability standards. This thorough verification process is essential for delivering a reliable, high-quality product to the market.

The verify phase validates that the design is robust and performs consistently under various conditions, addressing all identified risks and variabilities. Key objectives include:

1) **Validation of Design Performance:** Confirm through testing and analysis that the product meets specified performance and reliability criteria.

- Verify the product design meets robustness and reliability requirements aligned with consumer preferences from the Define phase.
- Confirm requirements at the component, sub-system, system, and vehicle levels based on project specifications.
- Use design verification methods (DVMs) to achieve design verification, incorporating significant Noise Factors.
- Confirm Reliable and Robust Product Function. Confirming reliable and robust product function involves:
  - Completing testing at all levels of the system's hierarchy
  - Completing reliability and robustness documentation
  - Completing design verification documentation
  - Completing testing at all levels of the system's hierarchy should be conducted as shown in Figure 4.18.
- Develop a plan to maintain product design robustness throughout its lifecycle.
- Assess system performance to ensure alignment with customer expectations during design and manufacturing, involving:
  - Evaluating Robustness
  - Conducting Design Review Based on Test Results (DRBTR)
  - Implementing Enhanced Testing Protocols
  - Creating a RDM
  - Analyzing Test Results
  - Verifying Manufacturing Processes
  - Updating d/pFMEA (Design and Process Failure Mode and Effects Analysis)
  - Implementing and Communicating Institutionalized Improvements

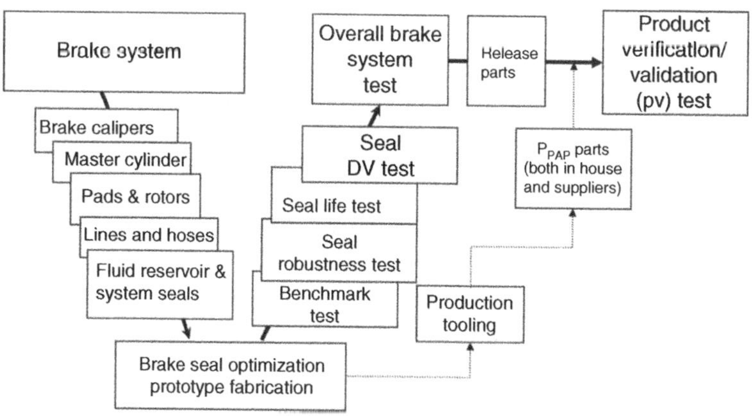

Figure 4.18 All Levels of System Hierarchy Tests.

2) Prototype Testing: Build and rigorously test prototypes to ensure they function correctly and consistently under expected conditions.
3) Reliability Testing: Conduct accelerated life testing and other reliability assessments to predict the product's lifespan and identify potential failure modes.
4) Field Testing: Deploy the product in real-world conditions to gather performance data and identify issues not evident in controlled testing.
5) Feedback Loop: Incorporate feedback from testing into the design, making necessary adjustments to improve performance and reliability.
6) Compliance and Standards Verification: Ensure the product complies with all relevant industry standards and regulatory requirements.
7) Final Risk Assessment: Re-evaluate any remaining risks and confirm they have been mitigated to an acceptable level.

For detailed information on objectives 1–7, many good references are available. Our featured approach strengthens the verify phase by focusing on these specific activities:

8) Robustness Assessment: Evaluate the product's ability to maintain performance under varying conditions and identify areas for robustness improvement.
9) Design Review Based on Failure Modes: Conduct thorough reviews to identify both soft and hard failure modes.
10) Final Design Review Based on Failure Modes: Perform a comprehensive final review to address any remaining failure modes.
11) Update Robustness and Reliability Demonstration Matrix: Reflect the latest test results and design changes to ensure all reliability and robustness targets are met.
12) Verify and Update CTQs/CTRs for Critical Components and/or Subsystems: Review and update critical-to-quality (CTQ) and critical-to-reliability (CTR) metrics for key components and subsystems, incorporating risk assessments to identify high-risk functions. Justify the need for PHM for these functions to ensure they meet necessary standards and contribute to overall product reliability.
13) Develop or Update Critical Parameter Management (CPM) Plan: Establish or revise the CPM plan to manage key parameters critical to the product's reliability and performance.

By thoroughly validating the product design, the verify phase ensures the final product is robust, reliable, and ready for market introduction. This comprehensive verification process minimizes the likelihood of post-launch failures, reduces warranty costs, and enhances customer satisfaction by delivering a product that performs reliably throughout its intended lifespan.

Note: The special featured approach for objectives 8–13 is enhanced in the activities section below in this chapter.

For potential questions about the difference between validation and verification, we provide the following validation and verification Comparison shown in Figure 4.19.

1) **Verification:**
   - **Purpose:** Ensures adherence to specifications and requirements (Voice of the Engineer [VOE]).

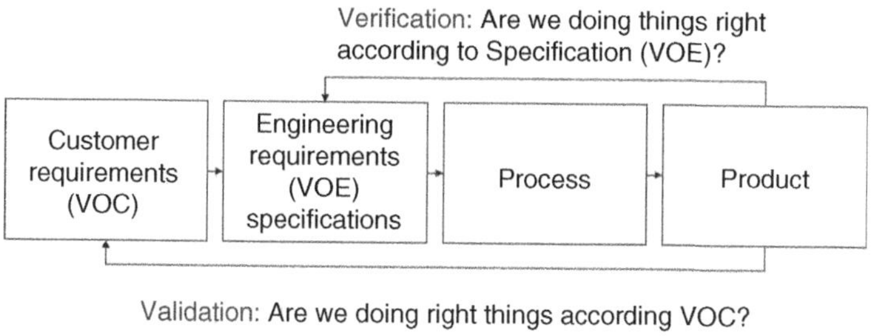

Figure 4.19 Validation vs Verification.

- **Focus:** Confirms that the product is being built correctly according to predefined standards and criteria.
- **Key Question:** Are we doing things right according to the specification (VOE)?
- **Examples:** Inspections, reviews, and testing during development to ensure compliance with design specifications.

2) **Validation:**
   - **Purpose:** Ensures that the product meets the customer's needs and expectations (Voice of the Customer [VOC]).
   - **Focus:** Confirms that the right product is being built to satisfy customer requirements.
   - **Key Question:** Are we doing the right things according to the VOC?
   - **Examples:** User acceptance testing, field trials, and feedback sessions to ensure the product fulfills customer needs and functions as intended in real-world conditions.

Verification is about ensuring the product is being developed correctly per technical specifications, while validation is about ensuring the final product meets the intended use and customer requirements. Both processes are crucial for delivering a high-quality, reliable product.

4.3.5.2 Verify Phase Activities in Robust Design for Reliability

The verify phase in RDfR is essential for ensuring that the product design meets all reliability and performance requirements before full-scale production begins. This phase aims to validate that the design is robust and performs consistently under various conditions, confirming that all identified risks and variability have been adequately addressed. While standard activities in this phase include validation of design performance, prototype testing, reliability testing, field testing, feedback loop integration, compliance and standards verification, and final risk assessment, our featured approach focuses on enhancing specific activities critical to ensuring robustness and reliability.

Key Activities are shown in Figure 4.20 in the verify phase.

By focusing on the 1–13 activities, the verify phase ensures that the final product design is robust, reliable, and ready for market introduction. This comprehensive verification process minimizes the likelihood of post-launch failures, reduces warranty costs, and enhances customer satisfaction by delivering a product that performs reliably throughout its intended lifespan.

Note: The detailed methodologies and tools for these key activities (8–13) are elaborated in the following sections in this chapter.

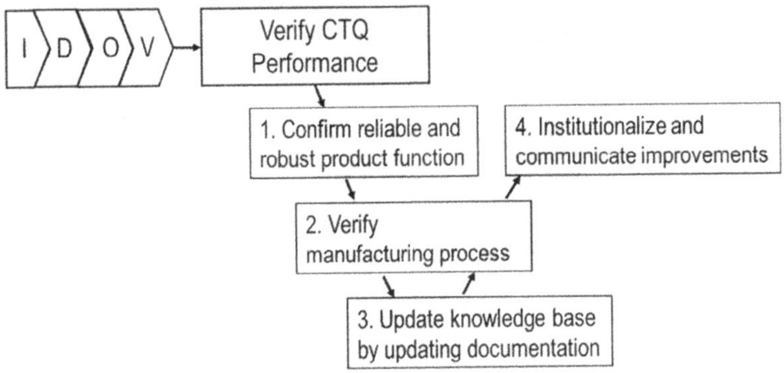

Figure 4.20 Verify Phase Activities Flow.

3) **Robustness Assessment:**
   - **Objective:** Evaluate the product's ability to maintain performance under varying conditions and identify areas for robustness improvement.
   - **Approach:** Conduct stress testing and environmental testing to simulate different operating conditions. Analyze test data to pinpoint weaknesses and implement design modifications to enhance robustness.

Robustness assessment is a strategy to *compare* the robustness of alternative systems [5]. This contrasts with parameter design, which is a strategy to *improve* robustness. Robustness assessment has great value, even after optimization using parameter design is done. It can also be used to:

- Validate design improvement based on different types of noise contributes to different rates and phases of failure as illustrated on a typical reliability "bathtub curve" shown in Figure 4.21.
- Compare new design to previous generation design as part of validation testing.
- Compare effectiveness of alternative test procedures.

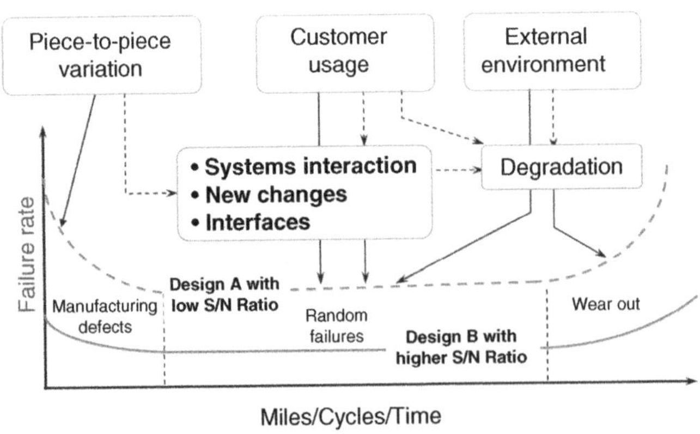

Figure 4.21 Noise Contributions to Reliability.

If the improved robustness can be validated, the potential failure modes will be minimized. After a parameter design optimization, it is recommended to do a robustness assessment of before and after improvements in the next subsystem/system to get confidence in a parameter design formulation (i.e., the response measure and the signal and noise conditions under which the response was measured).

To do robustness testing, select two products where one is already improved and better than the old and another one is the best in class (BIC) (i.e., a known improved good product and a known BIC). This testing determines if the data distinguishes between the two products and, therefore, builds confidence that the Parameter Design formulation is a good one.

The example below summarizes the "After Improvement, Before Improvement" strategy.

Table 4.3 Robustness Calculation Before and After Improvement.

	Robustness Calculation Before and After						
	Signal 1		Signal 2		Signal 3		
	N_1	N_2	N_1	N_2	N_1	N_2	
After Improvement (AI)	A_{11}	A_{12}	A_{21}	A_{22}	A_{31}	A_{32}	S/N_{AI}
Before Improvement (BI)	B_{11}	B_{12}	B_{21}	B_{22}	B_{31}	B_{32}	S/N_{BI}

The information contained in the robustness S/N ratio metric relates to the tendency of the system, or test procedure to give varying results. It should be clear at this point that robustness is a far more substantive evaluation of function than whether a system meets a target level of performance under a given set of test conditions. An example of testing two products is shown below. The first digit of the "y" subscript distinguishes the products as 1 or 2.

Table 4.4 Robustness Calculation Improved vs. BIC.

	Robustness Calculation Improved and BIC						
	Signal 1		Signal 2		Signal 3		
	N_1	N_2	N_1	N_2	N_1	N_2	
Improved	Y_{11}	Y_{12}	Y_{21}	Y_{22}	Y_{31}	Y_{32}	$S/N_{Improved}$
Best-in-Class (BIC)	BIC_{11}	BIC_{12}	BIC_{21}	BIC_{22}	BIC_{31}	BIC_{32}	S/N_{BIC}

- Verification test results should confirm that product performance meets customer expectations over the short and long term.
- Verification testing should be carried out at all levels of the system's hierarchy with emphasis placed on testing at the component and sub-system levels. Verification testing tests both the design of the product and the process that manufactures it.

Design Review Based on Failure Modes:

- **Objective:** Conduct thorough reviews to identify both soft and hard failure modes.
- **Approach:** Utilize FMEA to systematically review potential failure points. Engage cross-functional teams to assess and address soft (e.g., software-related) and hard (e.g., hardware-related) failure modes.

Design Review Based on Failure Mode:

- DRBFM is a systematic method of creatively preventing problems by focusing on the points of change, regardless of their origin. It is an extension of creative FMEA.
- DRBFM is best done with interdisciplinary teams in an environment that is conducive to sharing ideas freely.
- The goal of DRBFM is to maximize the reuse of robust components.

Team members should begin testing at the component level as soon as prototype parts are available. Since testing at all levels of the systems hierarchy is complex, it is important that this testing is well planned and managed. The DVP, also referenced in Optimize, is invaluable at this stage.

Analytical models often are used as the basis of verification testing because of the time and cost of hardware-based verification testing. For example, the transfer function is an ideal model that can be used both to predict product performance and verify the product design.

Final Design Review based on Failure Modes:

- **Objective:** Perform a comprehensive final review to address any remaining failure modes.
- **Approach:** Revisit the FMEA results and update with any new findings from recent tests. Ensure that all failure modes have been adequately mitigated or are within acceptable risk levels.

Update Robustness and Reliability Demonstration Matrix:

- **Objective:** Reflect the latest test results and design changes to ensure all reliability and robustness targets are met.
- **Approach:** Regularly update the robustness and RDM to capture the latest data. Use this matrix to communicate current reliability status and areas needing further improvement to the project team.

Reliability and robustness documentation should be updated with verification testing results.

It is equally important to simulate the stresses the sub-system and system are subjected to when in use; therefore, testing is conducted in the following way:

- The component is subjected to the stresses it will experience when part of a sub-system.
- Components are assembled into a sub-system and the sub-system is subjected to the stresses experienced when part of a system.
- The sub-system is assembled into a system, and the system is subjected to the stresses experienced when part of the whole product, e.g., vehicle.
- Finally, everything is assembled into the product system level and tested.

A Reliability and Robustness Demonstration Matrix (RRDM) is developed during the design phase to systematically organize and plan the verification activities required to demonstrate that a product satisfies its intended functional performance, reliability, and robustness requirements. As illustrated in Figure 4.22, the RRDM serves as a structured planning and tracking tool that links design risks and failure modes with specific verification methods and measurable outcomes. By doing so, it ensures that critical product characteristics identified during design analysis—such as those from DFMEA or robust design studies—are properly validated through targeted tests.

Reliability/Robustness Demonstration Matrix(RRDM)

System: Pump Motor	Assembly: Shaft Seal	Program	Date Originated:
Subsystem: N/A	Component(s): Seal	Latest Design Level	Date Revised:
		Lead Engineer:	

RDM Candidates			ROBUSTNESS IMPLEMENTATION							Demonstrated Results		
Key Function (s) & Rationale 1. Quality History(QH) 2. New Technology(NT) 3. DV Failure(DVF) 4. High Risks from DFMEA (HRFD))	High Impact Failure Modes	Relevant Components involved in failure	Verification Methods from RCL/DVP Including Test Name & Test ID if there is any.	*Output Response* Critical Function Metric(s): Units	Targets at Useful Life	Noise Factors Present in the Test: High Impact Noise Factors N1. Piece-to-Piece variation N2. Wear-out/drift over time N3. Customer Duty Cycle N4. Climate/Road conditions N5. System Interactions	Metric	Range	Noise Factors Implementation Strategies in Verification Methods	Demonstrated Validation Evidence & Comparison to Prior Design	Risk (R/ Y/ G) & Issues for Y or R.	Test Compl. Date
Example:												
Maintain pump pressure - pump rotor to housing / radial lip seal (This a carry over design with on going concern of seal.)	Shaft seal leaks at OD interface to housing	Pump shaft seal and pump housing	Seal Key Life Test KLT 1.3.1	PSI	Pressure decay of less than 12% per minute at useful life KLT (1000 test cycles)	N1. Housing Finish N5. Seal to housing press fit (N1. Seal OD - Housing ID) (Special parts will be built for test.)	um N mm	1-1.25 72 - 75 -0.15 to - 0.25	6 special parts will be built for high end (1.25 um) housing finish and low end housing ID (–0.25) with low seal to housing press fit (72N).	Baseline: 15% per minute pressure decay at 100 psi with 47.42 mm OD Seal in 47.32 mm housing with 1.95-2.05 um ID surface finish. Improved: 7% per minute at 100 psi with 47.52 mm OD in 47.32 mm housing, 1.15 um housing surface finish. 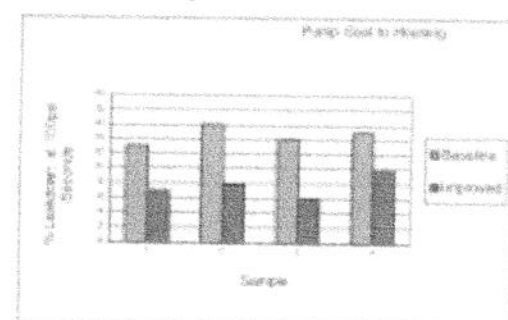	G (No issues)	2/8/05

Figure 4.22 Reliability and Robustness Demonstration Matrix.

The RRDM integrates both reliability verification and robustness evaluation within a single framework. Reliability verification focuses on demonstrating that the product performs consistently over time under expected operating conditions, while robustness evaluation verifies that the product continues to perform satisfactorily despite variations in environmental conditions, manufacturing tolerances, and noise factors. The matrix therefore provides a comprehensive view of how the design will be validated before release.

As shown in Figure 4.22, the RRDM is typically organized into several key sections. The first section identifies the system or subsystem under consideration, including the function, potential failure modes, and associated risks. This information often originates from design analyses such as DFMEA and helps ensure that verification efforts target the most critical risks.

The next section defines the verification requirements and robustness factors. These include the operating parameters, environmental conditions, and stress factors that the product must withstand, such as temperature ranges, vibration levels, loading conditions, or operational cycles. These factors are derived from customer requirements, engineering specifications, and the noise factors identified through robust design analysis.

Another major section of the matrix describes the planned verification tests. Each verification activity is clearly documented and typically includes:

- **Acceptance criteria:** Quantitative or qualitative performance thresholds that determine whether the test is considered successful. These criteria are based on engineering specifications, customer requirements, or regulatory standards.
- **Sample size:** The number of test units required to provide statistically meaningful results and sufficient confidence in the reliability assessment.
- **Test description:** A detailed explanation of the verification method, including the procedure, stress conditions, and measurement methods used to evaluate performance.
- **Test site:** The location where the verification will be conducted, such as a laboratory, proving ground, manufacturing facility, or field environment.
- **Responsibilities:** Identification of the engineering teams or individuals responsible for preparing, conducting, and documenting the test activities.
- **Test schedule or dates:** Planned timelines for executing the verification tests to ensure alignment with the overall product development schedule.

The RRDM also includes a results and evaluation section, where the outcomes of each test are documented. This section records the demonstrated results, compares them with the predefined acceptance criteria, and determines whether the product meets the required reliability and robustness targets. When results do not meet expectations, the matrix documents the corrective actions or design improvements that must be implemented. These actions may include design modifications, additional analysis, or repeated testing to verify that the issues have been resolved.

By consolidating all verification planning and outcomes into a single matrix, the RRDM provides clear traceability between design risks, verification activities, and final validation results. It also facilitates communication among design, testing, quality, and program management teams by providing a transparent overview of verification progress and remaining risks.

Ultimately, the RRDM plays a critical role in ensuring that the product design is thoroughly validated before production release. By systematically planning and documenting reliability and robustness verification, organizations can reduce the likelihood of field failures, improve product quality, and strengthen confidence that the design will perform reliably under real-world operating conditions.

Verify and Update CTQs/CTRs for Critical Components and/or Subsystems:

- **Objective:** Review and update CTQ and CTR metrics for key components and subsystems, incorporating risk assessments to identify high-risk functions.
- **Approach:** Conduct detailed risk assessments to identify high-risk functions. Justify the need for PHM for these functions to ensure they meet necessary standards. Update CTQs/CTRs accordingly to reflect these assessments and justifications.

The verification phase concludes once the team can confirm, using suitable robustness metrics, that both the design and manufacturing process can reliably meet CTQ or CTR requirements consistently over the intended lifespan, with minimal deviation from the target specifications.

- Achieve system- or subsystem-level CTQs/CTRs targets. As an example, shown in Figure 4.23.
- Minimize Variability in Y's
- In the brake case study, the robustness demonstration, as shown in Figure 4.24, enhanced torque across a range of hydraulic pressures. This significantly reduced audible noise and extended brake pad life.

Verify Design Recommendations Do Not Introduce Unintended Failure Modes: The RDFR team thoroughly examines proposed design changes or enhancements to ensure they do not inadvertently introduce new failure modes, such as overstress or weaknesses, into the system, as shown in Figure 4.25: Stress and Strength Analysis.

This process involves a comprehensive analysis of the potential impacts of the recommendations on the system's performance, reliability, and safety. It requires careful consideration of various factors, including the compatibility of the proposed changes with existing components, the system's operating conditions, and potential interactions between different subsystems. Additionally, it may involve conducting simulations, tests, or prototypes to validate the effectiveness and safety of the design recommendations before implementing them in the final product or system. Ultimately, the goal is to prevent any unintended consequences that could compromise the integrity or functionality of the system.

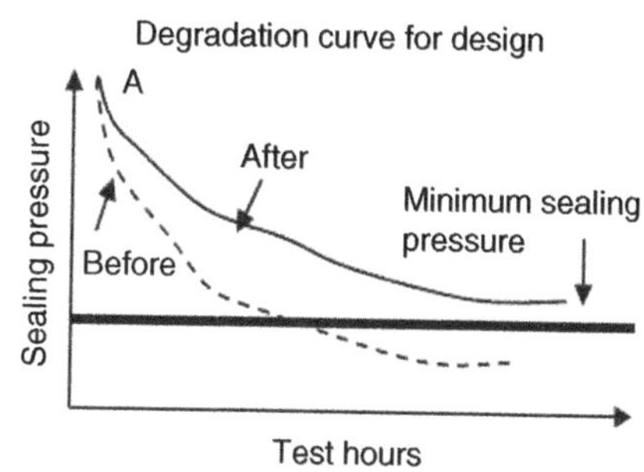

Figure 4.23 Graphical Demonstration for Degradation Curve.

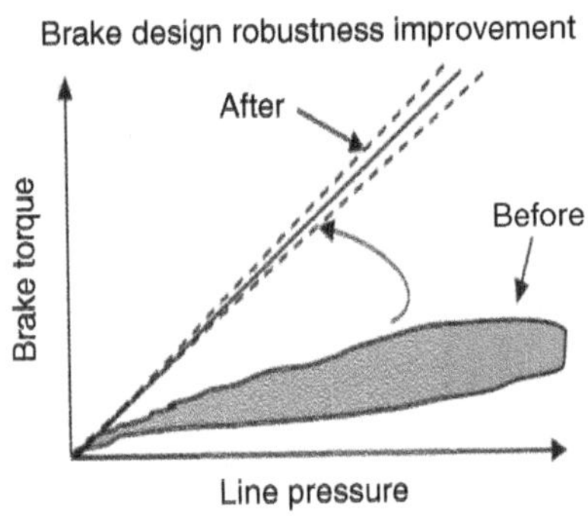

Figure 4.24 Example of Brake Design Robustness Improvement Improved Sensitivity and S/N Ratio.

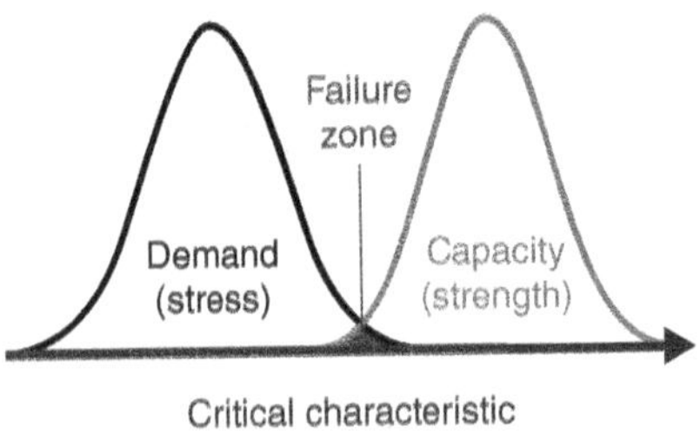

Figure 4.25 Stress and Strength Analysis Reduced Variation Significantly, Increased distance from failure mode.

Develop and Update Critical Parameter Management (CPM) Plan:

- **Objective:** Establish or revise the CPM plan to manage key parameters critical to the product's reliability and performance.
- **Approach:** Identify critical parameters that significantly impact reliability and performance. Develop or update the CPM plan to include monitoring and control strategies for these parameters, ensuring they stay within acceptable limits throughout the product lifecycle.

4.3.5.3 Verify Phase Deliverables

Deliverable	Responsibility Monitoring
Reliability Requirement	RDfR
Clear statement of use conditions	RDfR
Clear statement of lifecycle profile	RDfR
Summary of field reliability experience for similar products and components	RDfR
P-diagram for critical output response(s)	RDfR
DFMEA for the most critical risks	RDfR
Boundary diagram	RDfR
Noise Factor Management Strategy	RDfR
Reliability and Robustness Checklist	RDfR
Reliability and Robustness Demonstration Matrix	RDfR
Root Cause Analysis (RCA)/8D/DMAIC or equivalent	RDfR
Updated CPM plan	RDfR
PHM recommendations	RDfR

4.4 Robust Design Principles for Prognostic Health Management

Robust design is the proactive first step to enable and enhance the PoF for PHM development [2]. Robust design can play a crucial role as a proactive step in enabling and enhancing the PoF for PHM development. Using robust design principles for PHM development involves considering factors that can affect the reliability and performance of health indicators. The following is a step-by-step guide:

Define Objectives:

- Clearly define the objectives of your PHM system. Understand what health indicators you want to monitor and predict, identifying critical design parameters as the health indicators in earlier design not only assess the limitations of the given technologies adopted in the new design but also facilitate risk identification and mitigation, aiding decision-making, and budget planning for earlier investment.

Identification of Critical Parameters:

- Robust design involves the identification of critical parameters and factors that can influence the performance and reliability of a system. In the context of PHM, understanding the PoF requires a deep dive into the critical parameters affecting the health indicators.

Identify and select CTQs and CTRs as health indicators for PHM:

- In this section, we not only provide the "what" but also the "how" to identify the CTRs as the critical health indicators for PHM through the framework of identify, design, optimize, and verify (IDOV). This framework can be easily aligned with and adopted by businesses' needs for any reliability program, facilitating the design of reliability from robust birth reliability, tracking the health of products and equipment throughout their lifecycle, all while minimizing lifecycle costs.

At the heart of PHM with an RDfR approach lies the meticulous identification and selection of health indicators. These indicators serve as vital metrics, offering insights into system or component condition and performance. By carefully choosing and monitoring these indicators, organizations can gain early warnings of potential issues, enabling timely interventions and preventive maintenance actions [12].

- The significance of identifying and selecting health indicators cannot be overstated within the framework of effective PHM implementation. These indicators serve as the cornerstone, guiding decision-making processes and facilitating proactive maintenance strategies. Through a systematic approach to this process, organizations can enhance their ability to anticipate and mitigate potential failures, leading to improved reliability, increased uptime, and enhanced operational efficiency.

The specific benefits of implementing PHM, based on the identified health indicators through RDfR, include:

1) Capturing Product Functional and Environmental Conditions: PHM allows for the collection of data on product functional performance and environmental conditions, providing insights that contribute to a better understanding of mission requirements and

the definition of product specifications. This comprehensive understanding enhances the alignment of product design and performance with operational needs.

2) Improving Requirements for Validations and Verifications Testing: By leveraging PHM technology, organizations can enhance the requirements for validations and verifications testing. The data collected through PHM enables more accurate and targeted testing processes, ensuring that product performance meets or exceeds established standards and requirements.
3) Detecting and Preventing Product Failure: PHM facilitates early detection and prevention of potential product failures by continuously monitoring health indicators identified through RDfR. By analyzing trends and patterns in the collected data, organizations can proactively identify anomalies and potential failure modes, allowing for timely interventions and preventive maintenance actions to mitigate risks and prevent costly downtime.
4) Reducing Costs Associated with Downtime, Unscheduled Maintenance, and Spares Inventory: By proactively monitoring and managing product health, PHM helps minimize costs associated with downtime, unscheduled maintenance, and the maintenance of spare parts inventory. Early detection of issues and proactive maintenance interventions enable organizations to address potential problems before they escalate into costly failures, thereby reducing downtime and optimizing operational efficiency.

Overall, the implementation of PHM based on the identified health indicators through RDfR offers significant advantages in terms of enhancing product performance, tracking health of reliability, and cost-effectiveness across various stages of the product lifecycle. By leveraging PHM technology, organizations can gain valuable insights into product health, optimize maintenance strategies, and ultimately achieve improved operational efficiency and cost savings.

Robust Design for Reliability:

- The principles of robust design are inherently aligned with designing for reliability. By considering variations and uncertainties in operating conditions, materials, and other factors, robust design aims to create a system that is less sensitive to these variations, enhancing reliability.

Noise Factors and Control Factors:

- Robust design distinguishes between noise factors (uncontrollable sources of variation) and control factors (adjustable parameters). This differentiation is valuable in PHM as it allows for focusing efforts on mitigating the impact of noise factors while optimizing control factors for improved reliability.

Design Optimization:

- Optimization in robust design involves finding a design solution that minimizes the impact of variations and uncertainties. In the context of PHM, this optimization is critical for enhancing the PoF understanding and predictive capabilities.

Reliability Testing and Validation:

- Robust design encourages rigorous reliability testing to validate the system's performance across different conditions. This testing aligns with the goals of PHM development, where understanding and predicting failure modes based on the PoF are essential.

Proactive Approach:

- Robust design is inherently proactive. By addressing potential failure modes and variations during the design phase, it contributes to a proactive approach in PHM development, ensuring that the system is well prepared to handle and predict potential failures based on the underlying physics.

Documentation:

- Document all design decisions, parameters, and testing results. This documentation is crucial for future iterations, troubleshooting, and maintaining the PHM system.

Collaboration:

- Foster collaboration between engineers, data scientists, domain experts, and other stakeholders to ensure a comprehensive understanding of the health indicators and system dynamics.

Robust design in the context of PHM involves not only optimizing the system for accuracy but also making it resilient to uncertainties and variations that may occur in real-world scenarios.

Robust design acts as a foundational and proactive approach in PHM development by aligning with the principles of designing for reliability, understanding critical parameters, and optimizing the system's performance in the face of variations and uncertainties.

4.5 Scorecard for Robust Design for Reliability Implementation

Introduction

An RDfR implementation scorecard is a strategic tool used in product development to ensure that designs are reliable, robust, and capable of performing consistently under various conditions [13], including those influenced by uncontrollable environmental factors. This scorecard serves as a framework to assess and guide the reliability of engineering efforts throughout the design process, from initial concept to final product release. This approach not only enhances product reliability and customer satisfaction but also provides significant cost savings and a competitive edge in the market.

An RDfR scorecard offers substantial benefits by proactively addressing and desensitizing the harmful side effects of uncontrollable user environments. By using robustness as a leading indicator, organizations can drive reliability efforts and activities to the upfront and early design phase, enhancing product reliability, customer satisfaction. This approach not only enhances robust product performance but also provides cost efficiencies, competitive advantages, and effective risk management, leading to long-term success

Robust Design for Reliability (RDFR) Checklist

RDFR Scorecard Checklist	**Completed by:**		
Project:	**Date:**		
Product:	**Approved by:**	<Name, Function>	
Target Date:	**Date:**	<Latest Approval Date>	

IDOV Phase	Gate	Objective	Major Deliverables	Score	Actions Required	Responsible Party	Due Date	Notes
Identify (I)	Gate 1: Avoid failure modes due to bad planning and definition	Ensure all customer requirements, potential risks, and operationa conditions are captured early in the design process.	- Quality History: Analyze past product issues and successes to avoid similar issues.					
			- Front Load Challenges & New Technologies: Identify challenges and new technologies early in the design process.					
			- NUD (New, Unique, Difficult): Define new, unique, and difficult aspects of the project.					
			- Kano Model: Use Kano analysis to prioritize customer requirements.					
			- Functional Targets: Set clear functional and performance targets.					
			- Reliability Requirements: Establish reliability objectives based on customer expectations.					
			- Operational Noise: Identify environmental or operational conditions that could affect reliability.					
			Capture Reliability Lessons Learned					
			- PHM Assessment: Identify the PHM requirements based on system design and customer needs and lesson learned.					
	Gate 2: Avoid failure modes due to poor formulation and characterization	Ensure clear problem formulation and proper characterization of the system to support reliable design solutions.	- Transfer Functions: Establish relationships between design parameters and performance.					
			- Parsimonious Bill of Materials (BOM): Optimize the BOM to minimize unnecessary complexity.					
			- P-Diagrams: Map the relationships between input, output, control, and noise factors.					
			- FMEAs: Conduct Failure Modes and Effects Analysis (FMEA) to identify potential risks.					
			- Robustness Checklist (RCL): Create a checklist to ensure robustness considerations are covered.					
			- Initial Noise Factor Management Strategy: Outline a strategy for managing noise factors early in the design.					
			Prognostic Health Management (PHM) Needs: Determine the Needs for PHM during product operation.					

Figure 4.26 Customers' Need Review and RDFR Team Launch.

and market leadership. Furthermore, it presents a significant opportunity to identify and select meaningful health indicators for PHM applications, thereby ensuring comprehensive lifecycle reliability management.

Key Components of the Scorecard as shown in following figures:

Figure 4.26: RDFR Team Launch and Customers' Need Review

Figure 4.27: Concept Review

Figure 4.28: Preliminary Design Review

Figure 4.29: Final Design and Verify Review

1) Design Objectives and Requirements
   - Critical-to-Quality (CTQ) and Critical-to-Reliability (CTR) Parameters: Identifies key performance indicators that must be met for the design to be considered robust and reliable.
   - Customer Requirements: Ensures that the design meets the needs and expectations of the end-users.

Robust Design for Reliability (RDFR) Checklist

RDFR Scorecard Checklist	**Completed by:**
Project:	**Date:**
Product:	**Approved by:** <Name, Function>
Target Date:	**Date:** <Latest Approval Date>

IDOV Phase	Gate	Objective	Major Deliverables	Score	Actions Required	Responsible Party	Due Date	Notes
Design (D)	Gate 3: Avoid failure modes due to poor technical execution	Ensure robust technical execution to prevent the introduction of new failure modes during the design phase.	- **Parameter Studies**: Analyze how design parameters affect system performance and robustness.					
			- **RSOW (Reliability Statement of Work)**: Define reliability objectives and criteria.					
			- **Failure Mode – Noise Factor Mapping**: Map failure modes to noise factors that could influence them.					
			- **Robustness Plan**: Develop a comprehensive robustness plan to address potential weaknesses.					
			- **Updated Noise Factor Management Strategy**: Refine strategies for managing noise factors based on updated data.					
	Mistake Prevention Plan	Proactively prevent mistakes in design and manufacturing processes.	- **DFMEA (Design FMEA)**: Identify and address potential failure modes in the design.					
			- **PFMEA (Process FMEA)**: Identify potential risks in the manufacturing process.					
			- **SDS (System Design Specification) Deviations**: Manage deviations from design specifications.					
			- **Design Reviews**: Conduct reviews to validate key aspects of the design.					
			- **CPM (Critical Parameter Management) Plan**: Identify and manage critical parameters.					
			- **DRBFM (Design Review Based on Failure Modes)**: Use failure modes to guide detailed design reviews.					

Figure 4.27 Design Phase.

2) Risk Assessment and Mitigation
   - Failure Modes and Effects Analysis (FMEA): Identifies potential failure modes and their effects on the system, enabling proactive mitigation.
   - Risk Prioritization: Ranks identified risks based on their severity, occurrence, and detectability to focus on the most critical issues.

3) Design Verification and Validation
   - Prototyping and Testing: Involves creating prototypes and conducting rigorous tests to validate that the design meets reliability standards.
   - Design Verification Plan and Report (DVP&R): Documents the testing procedures and results to ensure thorough verification.

4) Robustness Assessment
   - Stress and Strength Analysis: Evaluates the product's ability to withstand different stresses without failure.
   - Environmental Testing: Assesses the product's performance under various environmental conditions (e.g., temperature, humidity, vibration).

5) Continuous Improvement and Feedback
   - Lessons Learned: Incorporates feedback from testing and real-world use to continuously improve the design.

Robust Design for Reliability (RDFR) Checklist

RDFR Scorecard Checklist	**Completed by:**		
Project:	**Date:**		
Product:	**Approved by:**	<Name, Function>	
Target Date:	**Date:**	<Latest Approval Date>	

IDOV Phase	Gate	Objective	Major Deliverables	Score	Actions Required	Responsible Party	Due Date	Notes
Optimize (O)	Gate 4: Avoid failure modes due to lack of robustness and mistakes (hard and soft failures)	Optimize the design to withstand uncontrollable user environments and mitigate both hard (physical) and soft (functional) failures. Prevent misakes in design, manufacturing, and operation.	- **Robustness Design Strategy**: Ensure that the product is robust against external noise factors (e.g., temperature, humidity, user behavior). Design the product to be insensitive to such environmental variances while maintaining functionality.					
			- **Producibility**: Ensure the design is optimized for manufacturing and assembly, reducing the likelihood of production errors.					
			- **Mistake Prevention**: Apply mistake-proofing (poka-yoke) techniques to eliminate operational and manufacturing errors.					
			- **Environmental Noise Management**: Implement design elements that address challenges posed by uncontrollable external noise factors.					
			- **Design Verification Plan (DVP)**: Establish a detailed plan for testing and validating the design under real-world conditions.					
			- **Reliability Demonstration Matrix (RDM)**: Track reliability metrics to ensure the design meets reliability objectives.					
			- **Hard Failure Mitigation**: Ensure design robustness to prevent material and mechanical breakdowns.					
			- **Soft Failure Mitigation**: Ensure design robustness to prevent performance degradation due to operational variations.					

Figure 4.28 Optimize Phase.

- Design Reviews: Conducts regular reviews with cross-functional teams to ensure alignment and address any emerging issues.

Benefits of the Scorecard

1) Enhanced Product Reliability
   - Ensures that products perform consistently and reliably, even under challenging conditions.
   - Reduces the likelihood of failures, leading to increased customer satisfaction and trust.

2) Cost Savings
   - Minimizes warranty claims and maintenance costs by reducing field failures.
   - Optimizes resource allocation by focusing on critical design aspects, reducing rework and scrap.

3) Proactive Risk Management
   - Identifies and addresses potential failure modes early in the design process, mitigating risks before they become costly issues.
   - Facilitates the implementation of robust design principles that reduce sensitivity to environmental noise and variability.

Robust Design for Reliability (RDFR) Checklist

RDFR Scorecard Checklist	**Completed by:**
Project:	**Date:**
Product:	**Approved by:** <Name, Function>
Target Date:	**Date:** <Latest Approval Date>

IDOV Phase	Gate	Objective	Major Deliverables	Score	Actions Required	Responsible Party	Due Date	Notes
Verify (V)	Gate 5: Validation & Deployment - Avoid failure modes due to unrepresentative tests	Ensure the product is properly tested for robustness and mistake-proofing. Validation ensures the product can perform under all expected conditions.	- Design Verification Report: Document final testing results, including results from upgraded tests that reflect real-world use cases.					
			- Reliability Demonstration Matrix (RDM): Use the RDM to ensure the product has been tested against all potential failure modes.					
			- Robustness Assessment: Evaluate the product's ability to maintain performance under varying conditions and identify areas for robustness improvement.					
			- Design Review based on Failure Modes (DRBFM): Conduct thorough reviews to identify both soft and hard failure modes.					
			- Final Design Review based on Test Results (DRBTR): Perform a comprehensive final review to address any remaining failure modes.					
			- Update Robustness and Reliability Demonstration Matrix: Reflect the latest test results and design changes to ensure all reliability and robustness targets are met.					
			- Verify and Update CTQs/CTRs for Critical Components and/or Subsystems: Review and update Critical to Quality (CTQ) and Critical to Reliability (CTR) metrics for key components and subsystems, incorporating risk assessments to identify high-risk functions.					
			-Develop or Update Critical Parameter Management (CPM) Plan: Establish or revise the CPM plan to manage key parameters critical to the product's reliability and performance.					
			- Identify and document Health Indicators for PHM based on CTQ and CTR					
			- Capture Reliability Lessons Learned					

Figure 4.29 Final Design and Verify Review.

4) Competitive Advantage
   - Produces high-quality, reliable products that stand out in the market.
   - Enhances the brand's reputation for reliability and robustness, attracting more customers.
5) Lifecycle Reliability Management
   - Supports PHM by identifying meaningful health indicators that provide early warnings of potential failures.
   - Ensures comprehensive reliability management throughout the product lifecycle, from development to end of life.

Implementation Steps
Please see figures below for details
Figure 4.26: RDFR Team Launch and Customers' Identify Phase
Figure 4.27: Design Phase
Figure 4.28: Optimize Phase
Figure 4.29: Verify Review

4.6 Digital Quality Through Robust Design for Reliability

Digital quality is an increasingly important aspect of RDfR, especially as products become more complex and interconnected [14]. In the context of RDfR, digital quality encompasses the use of advanced digital technologies and data-driven approaches to enhance product reliability, performance, and overall quality. Let's elaborate on the key aspects of digital quality in RDfR:

Advanced Data Analytics:

- Implement big data analytics to identify patterns and predict potential reliability issues.
- Use machine learning algorithms to optimize design and manufacturing processes.
- Develop real-time monitoring systems for product performance and quality.

Expansion:

- Predictive Analytics: Utilize historical data and real-time inputs to forecast potential failures or performance degradation.
- Anomaly Detection: Implement AI-driven systems to identify unusual patterns that may indicate impending failures.
- Process Optimization: Use data-driven insights to continuously refine manufacturing processes for improved reliability.

Digital Twin Technology:

- Create digital twins of products to simulate performance under various conditions.
- Use digital twins for predictive maintenance and reliability optimization.
- Integrate digital twin data with product lifecycle management systems.

Expansion:

- Virtual Testing: Conduct extensive virtual tests on digital twins to identify potential failure modes without physical prototyping.
- Real-time Performance Monitoring: Use digital twins to compare actual product performance with expected behavior.
- Design Iteration: Rapidly iterate designs using insights from digital twin simulations.

Internet of Things (IoT) and Connected Products:

- Implement IoT technologies to gather real-time data on product performance.
- Use connected product data to inform future design improvements.
- Develop predictive maintenance capabilities based on IoT data.
- Edge Computing: Implement edge computing solutions for real-time data processing and rapid response to potential issues.
- Ecosystem Integration: Design products to integrate seamlessly with other connected devices and systems.
- User Behavior Analysis: Analyze data from connected products to understand how users interact with the product in real-world scenarios.

Automated Quality Control:

- Implement AI-powered visual inspection systems.
- Develop automated testing and validation processes.
- Use robotics and automation to ensure consistent quality in manufacturing.

Expansion:

- Machine Vision: Utilize advanced image processing and AI to detect defects that might be missed by human inspectors.
- Autonomous Testing: Develop self-running test suites that can adapt to new product variations or identified issues.
- Continuous Monitoring: Implement systems for ongoing quality checks throughout the manufacturing process.

Blockchain for Quality Assurance:

- Implement blockchain technology for secure and transparent quality records.
- Use smart contracts to automate quality control processes.
- Ensure the integrity and traceability of quality-related data throughout the product lifecycle.
- Supply Chain Transparency: Use blockchain to track components and materials from source to final product.
- Automated Compliance: Implement smart contracts that automatically verify compliance with quality standards and regulations.
- Immutable Quality Records: Create tamper-proof records of quality tests, inspections, and certifications.

Digital Simulation and Modeling:

- Utilize advanced simulation software to model product behavior under various conditions.
- Implement multi-physics simulations to understand complex interactions within products.
- Use AI-enhanced modeling to predict long-term reliability and performance.
- Scenario Planning: Model a wide range of potential use cases and environmental conditions to ensure robust design.
- Material Science Integration: Incorporate advanced material models into simulations for more accurate predictions of long-term behavior.
- System-level Simulations: Model entire systems or ecosystems to understand how products interact with other components or products

Augmented and Virtual Reality in Design and Quality Assurance:

- Use AR/VR technologies for virtual prototyping and design reviews.
- Implement AR-assisted assembly and quality inspection processes.
- Develop VR training programs for quality control and maintenance procedures.
- Collaborative Design: Use VR environments for global teams to collaborate on design reviews and problem solving.
- Guided Assembly: Implement AR systems to guide workers through complex assembly processes, reducing errors.
- Virtual User Testing: Conduct user experience tests in VR environments to identify potential usability issues early in the design process

Cybersecurity in Product Design:

- Integrate cybersecurity considerations into the RDfR process from the outset.
- Implement secure-by-design principles in connected products.
- Develop processes for ongoing security updates and vulnerability management.
- Threat Modeling: Conduct comprehensive cybersecurity threat modeling as part of the design process.
- Secure Communications: Design robust, encrypted communication protocols for IoT and connected products.
- Automated Security Testing: Implement continuous security testing processes to identify and address vulnerabilities throughout the product lifecycle.

By integrating these digital quality aspects into the RDfR process, organizations can significantly enhance their ability to design, manufacture, and maintain highly reliable products. This approach not only improves product quality and reliability but also provides valuable insights for continuous improvement and innovation. As digital technologies continue to evolve, their integration into RDfR practices will become increasingly crucial for maintaining competitiveness in the global market.

Implementing Digital Quality Through Robust Design for Reliability (RDfR)

Digital quality in the context of RDfR is about leveraging digital tools, techniques, and processes to ensure that products are designed and manufactured to meet high standards of reliability and performance. This involves integrating advanced digital technologies into the RDfR framework to enhance the quality and reliability of products throughout their lifecycle.

Key Aspects of Digital Quality Implementation

1) Data-Driven Decision-Making
   - **Big Data and Analytics:** Utilize big data analytics to gather and analyze vast amounts of data from various stages of the product lifecycle. This includes data from design, manufacturing, testing, and field performance.
   - **Predictive Analytics:** Implement predictive analytics to forecast potential failure modes and reliability issues before they occur. This proactive approach allows for timely interventions and design modifications.
2) Simulation and Modeling
   - **Computer-Aided Engineering (CAE):** Use CAE tools to simulate and model the behavior of products under various conditions. This helps in identifying potential failure points and optimizing the design for robustness and reliability.
   - **Digital Twins:** Develop digital twins of products to simulate real-world performance in a virtual environment. This allows for continuous monitoring and improvement of product designs based on real-time data.
3) Advanced Testing and Validation
   - **Virtual Testing:** Conduct virtual tests to evaluate product performance under different scenarios without the need for physical prototypes. This accelerates the testing process and reduces costs.
   - **Accelerated Life Testing (ALT):** Perform accelerated life testing to identify failure modes and mechanisms quickly. This helps in understanding the long-term reliability of products in a shorter timeframe.

4) Quality Management Systems (QMS)
   - **Automated Quality Control:** Implement automated quality control systems to monitor and control the quality of products during manufacturing. This ensures consistent adherence to quality standards.
   - **Real-Time Quality Monitoring:** Use real-time monitoring systems to track quality metrics and parameters throughout the production process. Immediate feedback allows for quick corrective actions.
5) Collaborative Platforms
   - **Integrated Design Environments:** Utilize collaborative platforms that integrate design, engineering, and manufacturing teams. This fosters real-time collaboration and ensures that all stakeholders are aligned on quality and reliability goals.
   - **Cloud-Based Solutions:** Adopt cloud-based solutions to store and share data across the organization. This enhances accessibility and ensures that everyone has the latest information.

Steps for Implementing Digital Quality Through RDfR

1) Define Quality and Reliability Objectives
   - Establish clear quality and reliability objectives based on customer requirements and industry standards.
   - Develop key performance indicators (KPIs) to measure and track the achievement of these objectives.
2) Leverage Digital Tools and Technologies
   - Identify and implement appropriate digital tools and technologies that support RDfR, such as CAD/CAE software, simulation tools, and quality management systems.
   - Train employees on the use of these tools to ensure effective adoption and utilization.
3) Integrate Digital Quality into the Product Development Process
   - Incorporate digital quality practices at every stage of the PDP, from concept to production.
   - Use digital simulations and models to evaluate design alternatives and optimize for reliability.
4) Implement Continuous Monitoring and Improvement
   - Establish systems for continuous monitoring of product quality and performance using real-time data.
   - Conduct regular reviews and audits to identify areas for improvement and implement corrective actions.
5) Foster a Culture of Quality and Reliability
   - Promote a culture that values quality and reliability across the organization.
   - Encourage collaboration and communication among teams to ensure a unified approach to digital quality implementation.

Benefits of Digital Quality Implementation Through RDfR

- **Enhanced Product Reliability:** By leveraging digital tools and technologies, organizations can design and manufacture products that are more reliable and perform consistently under various conditions.

- **Reduced Time to Market:** Digital simulations and virtual testing accelerate the PDP, allowing for quicker identification and resolution of potential issues.
- **Cost Savings:** Reducing the need for physical prototypes and manual quality checks lowers costs associated with product development and testing.
- **Improved Customer Satisfaction:** High-quality, reliable products lead to increased customer satisfaction and loyalty.
- **Data-driven Insights:** Real-time data and analytics provide valuable insights that can be used to continuously improve product designs and processes.

Implementing digital quality through RDfR involves integrating advanced digital tools and techniques into the design and manufacturing processes. By leveraging data-driven decision-making, simulation and modeling, advanced testing, quality management systems, and collaborative platforms, organizations can enhance the quality and reliability of their products. This comprehensive approach not only improves product performance and customer satisfaction but also drives innovation and reduces time-to-market and costs.

4.7 Critical Parameter Development and Management (CPD&M) Process and Phases

Critical Parameter Development and Management (CPD&M) is a comprehensive and systematic approach to product development and lifecycle management that focuses on identifying, analyzing, and controlling the most crucial parameters affecting a product's performance, reliability, and quality [1]. This process can be integrated into the broader product development cycle or implemented as a standalone initiative for post-launch improvements.

At its core, CPD&M is a collaborative effort involving a cross-functional team of engineers and technical specialists who work together to uncover and manage the parameters that govern the functionality of any engineered system. This approach aligns closely with Systems Engineering principles, emphasizing a holistic view of product development and management.

The CPD&M process begins with a thorough analysis of all design parameters. From this comprehensive set, the team identifies and isolates those parameters that pose the highest risk to the product's performance – these are designated as critical parameters. The primary objective is to improve or control these critical parameters throughout the entire product lifecycle, from initial design to manufacturing, packaging, and maintenance.

The CPD&M process can be broken down into 12 steps across 5 phases:

Phase 1: Planning and Requirements Assessment

1) Create a Project Charter
2) Assemble a cross-functional expert team
3) Assess product requirements

Phase 2: Diagramming and Parameter Identification

4) Construct Function Trees and Flow Diagrams
5) Develop various diagrams (I-O-C, P-Diagrams, Noise Diagrams, etc.)

Phase 3: Critical Parameter Analysis

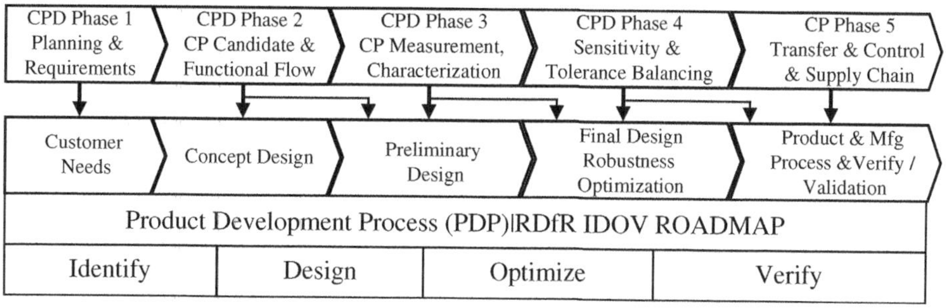

Figure 4.30 CPD&M Phase Alignment with the Product Development Process.

6) Conduct CP measurement and experimentation
7) Ensure measurement system capability
8) Design and conduct experiments
9) Analyze data using statistical methods
   Phase 4: Tolerance Establishment
10) Verify tolerance ranges and variation contributions
   Phase 5: Implementation and Control
11) Create a manufacturing and production CP implementation plan
12) Develop an ongoing control and evaluation plan

CPD&M Phase Alignment with the Product Development Process

When proactively applying CPD&M during Technology Development or Product Development, the following illustration shows how the Phases align to Product Development Phases (Figure 4.30):

One of the key outputs of the CPD&M process is the creation of a Design Guide. This document serves as a comprehensive reference that explains the fundamental workings of the engineered system, detailing how and why it functions as it does, and highlighting its sensitivities. This guide becomes an invaluable resource for current and future development efforts, ensuring that critical knowledge is preserved and easily accessible.

CPD&M is a vital process within Systems Engineering that ensures the identification and control of parameters crucial to the performance and reliability of engineered systems. By integrating CPD&M into the PDP and maintaining rigorous post-launch management, companies can mitigate risks, enhance product quality, and foster cross-functional collaboration, ultimately leading to greater customer satisfaction and business success.

References

1 Ambler, A.P. and Sheppard, J.W. (ed.) (2025). *Realizing Complex Integrated Systems.* Taylor & Francis Group.

2 Physics-of-Failure: An Approach to Reliable Product Development by Michael Pecht and Abhijit Dasgupta, 95 IRW Final Report 1995.

3 Clausing D. (1994). Total Quality Development. *A Step-By-Step Guide to World-Class Concurrent Engineering*, New York: ASME Press.

4 Readiness Plan, Transfer function-based design to improve product reliability and robustness in design for six sigma, Matthew Hu, Six Sigma Forum Magazine, American Society for Quality, August 2013.

5 Clausing, D. and Frey, D.D. (ed.) (2006). *Improving System Reliability by Failure Mode Avoidance Including Four Concept Design Strategies.* Massachusetts, USA: Wiley InterScience.

6 Fowlkey, W.Y. and Creveling, C.M. (1995). *Engineering Methods for Robust Product Design.* Addison-Wesley Publishing Company. Massachusetts, USA: Reading.

7 Taguchi, G., Chowdhury, S., and Taguchi, S. (2000). *Robust Engineering, Learn How to Boost Quality While Reducing Costs & Time to Market.* New York, NY, USA: McGraw Hill.

8 Taguchi, G., Chowdhury, S., and Wu, Y. (2004). *Taguchi Quality Engineering Handbook.* Hoboken, New Jersey, USA: Wiley.

9 Taguchi, G. and Yokoyama, Y. (1993). *Taguchi Methods: Design of Experiments.* American Supplier Institute.

10 Phadke, M.S. (1989). *Quality Engineering Using Robust Design.* Dearborn, Michigan, USA: Prentice-Hall.

11 Matthew, H. and Yang, K. (2005). Transfer function development in design for six sigma framework. SAE Technical Paper 2005-01-1215. *Society of Automotive Engineers.*

12 Liao, L. and Siegel, D.A. (2014). *Mechanical Systems and Signal Processing.*

13 Miller, R.J., Green, J., Herrmann, D., and Heer, D. (2004). *Assess Your Program for Probability of Success using the Reliability Scorecard Tool.* RAMS: IEEE.

14 Yang, K. (2024). *Quality In the era of Industry 4.0.* Wiley.

Further Reading

Davis, T.P. (2006). Science, engineering, and statistics. *Applied Stochastic Models in Business and Industry* 22: 401–430.

Otto, K. and Wood, K. (2001). *Product Design, Techniques in Reverse Engineering and New Product Development*, Prentice Hall.

5

Predictive & Health Management

CHAPTER MENU

5.1 Justification for PHM in Robust System Design

In today's unpredictable and complex environments, designing reliable and robust systems requires more than just preventing failure. Prognostic health management (PHM) plays a key role by helping engineers detect problems early, monitor system health in real time, and make informed decisions throughout the design and operation of a system.

This chapter explains why PHM should be included early in the system design process. It shows how PHM can strengthen engineering confidence, improve reliability, and reduce the risk of failure—even when systems are exposed to changing or uncontrollable conditions. Unlike traditional failure analysis, which reacts after something goes wrong, PHM focuses on early warnings and preventing issues before they occur.

By combining predictive tools and real-world monitoring, PHM supports better design decisions, lowers maintenance costs, and increases system performance and safety. Overall, PHM is presented as a valuable approach for building systems that are both reliable and resilient in real-world use.

Investing in PHM is not just about avoiding failures—it is about reinforcing engineering intent, ensuring design robustness, and future-proofing industrial systems in an increasingly complex and unpredictable operational landscape.

In modern engineering, **ensuring reliability, efficiency, and safety in complex systems begins at the upfront design phase**. **Engineering confidence** is established not only through design validation but also by proactively understanding the **limitations of products and technologies** under **uncontrollable user environments**. **PHM** is not merely a **failure detection mechanism** but an **essential component of robust system design** that strengthens **engineering confidence** by ensuring that a system performs **as intended despite variations in real-world conditions**.

Engineering Confidence Through Robust Design

Robust engineering focuses on designing systems that consistently meet performance objectives **despite environmental and operational uncertainties**. To build such confidence in performance, PHM must be integrated early in the design process. It plays a critical role in:

- **Identifying key design parameters as potential health indicators** to assess system behavior under different conditions.
- **Quantifying and managing uncertainties** to minimize the risk of unexpected failures.
- **Ensuring that deviations from intended design performance are detected, predicted, and mitigated** before they escalate into failure states.

Table 5.1 Traditional Failure Analysis vs. PHM.

	Traditional Failure Analysis	Prognostic Health Management in Early Design Phases
Purpose	Investigate the cause of failure after it occurs	Evaluate system performance under variation to prevent failure
Timing	Reactive—conducted after a failure	Proactive—conducted before failure, during design or testing
Failure Detection	Post-failure analysis	Early detection of degradation trends and performance shifts
System Monitoring	Limited or no continuous monitoring	Continuous monitoring
Decision-Making	Based on past failures and incidents	Based on predictive analytics and real-time health indicators
Focus	Identifies root cause of failure and mitigate failure modes	Examines sensitivity to noise factors and variability. Ensure design intent is met under varying conditions

Source: Vogl et al. [1] / The Prognostics and Health Management Society / CC by 3.0.

Unlike traditional failure analysis, which often reacts to failure **after it occurs**, PHM aligns with **design intent** by proactively:

- **Imposing uncontrollable environmental variables** into the design process to anticipate real-world impacts.
- **Identifying degradation trends and performance shifts early** to enable timely corrective actions.
- **Minimizing costly over-design** by strategically balancing **performance, cost, and reliability**.

Building upon the concept of engineering confidence, it is essential to consider design limitations in uncontrollable user environments.

Understanding Design Limitations in Uncontrollable User Environments

No design exists in isolation—systems must perform **consistently** under **dynamic and often unpredictable external factors** such as:

- Varying temperature, humidity, and pressure conditions in aerospace applications.
- Unstable supply chain and manufacturing variability in industrial systems.
- Heavy loads, vibrations, and stress variations in automotive engineering.
- Complex patient interactions and medical uncertainties in healthcare technology.

PHM provides **continuous insight into system behavior** under these **real-world operating conditions**, enabling engineers to:

- **Refine design tolerances** by evaluating **how key parameters influence system reliability**.
- **Assess risks early** and make data-driven decisions on **the need for PHM** to maintain system health.
- **Validate robustness** by ensuring that the design intent remains **intact despite external variations**.

By embedding **PHM into the design process**, engineers gain **empirical data on operational behavior**, ensuring that potential **failure risks are anticipated rather than reacted to**. Moreover, PHM bridges the gap between controlled design assumptions and messy real-world conditions, offering feedback loops that inform both design and operation.

Failure Modes as Deviations from Intended Design Performance

In **robust engineering**, failure is **not the design intent**—it is an **unintended deviation** from **expected functional performance** when subjected to external stressors (Figure 5.1).

Traditional failure analysis often misplaces focus on **mitigating failure modes** rather than **understanding the root causes of deviation** from intended design function. PHM reframes this by:

- **Detecting early warning signs of deviation** before they compromise system integrity.
- **Ensuring that corrections** are made proactively to sustain normal system performance.
- **Utilizing key health indicators** to optimize system resilience and extend asset lifespan.

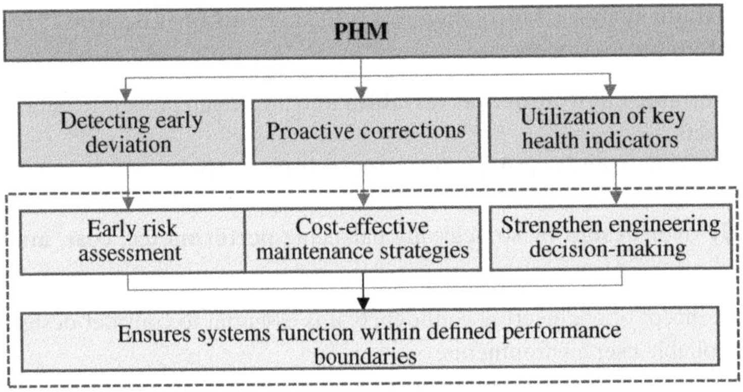

Figure 5.1 PHM Help Ensures Design Intent Is Met.

Rather than defining success as **merely avoiding failure**, **PHM-driven design confidence** ensures that systems function **within defined performance boundaries**, allowing engineers to:

- Optimize product resilience through early risk assessment.
- Develop cost-effective maintenance strategies to prevent downtime.
- Strengthen engineering decision-making by linking design parameters with reliability outcomes.

Summary: PHM as an Enabler of Intentional, Resilient Design

PHM is **not just a tool for failure prevention**—it is a **strategic enabler of robust design intent** that ensures:

- Early identification of critical design parameters as health indicators to evaluate the risk of system failures.
- Engineering confidence through real-world stress testing in uncontrollable user environments.
- Proactive system resilience strategies that optimize performance, cost, and reliability.

By integrating predictive analytics, real-world condition monitoring, and data-driven insights, PHM transforms system design from a reactive discipline into a proactive strategy, ensuring that engineering confidence is upheld even in the face of unpredictable real-world variations.

5.2 System Components and Their Functions

5.2.1 PHM System Architecture

The fault diagnosis and prognosis mainly involve condition monitoring, data collection, fault diagnosis, fault prognosis, and maintenance decision-making for key systems such as the mechanical system, software system, and electrical system. Fault diagnosis and prognosis should be combined with the operation system of the product and can also be used in independent systems.

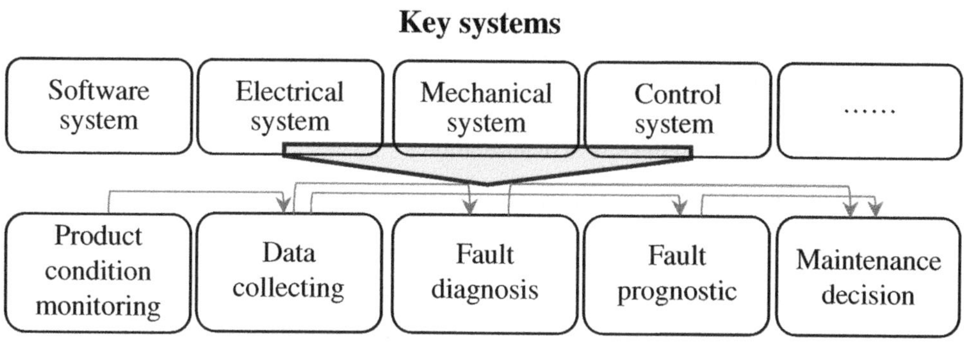

Figure 5.2 Overall Framework of PHM System.

Figure 5.2 is a schematic diagram of PHM system. Its operation process should include five main steps: product condition monitoring, data collection, fault diagnosis, fault prognosis, and maintenance decision-making.

Product Condition Monitoring: Product condition monitoring should select:

- Appropriate measurement techniques.
- Measurement locations.
- Sensors to measure certain parameters that require measurement, achieving the monitoring function of the system status.

The layout of sensors should consider principles of economy, fault detectability, fault identifiability, and minimal impact on the original product.

Monitoring parameters: According to the preselected monitoring plan, there may be one or several measurement techniques suitable for specific measurable parameters. Measurement result parameters may be obtained through direct or indirect measurement methods. For some parameters (such as current, voltage, and vibration), simple measurement values may not be sufficient to indicate the occurrence of a fault. Auxiliary methods may be required to reveal the changes caused by the fault.

Monitoring locations: Monitoring locations should be selected where faults can be detected. Measurement points should be uniquely marked, preferably with permanent signs or identification codes. Factors to consider include operational safety, sensor selection, signal processing, high sensitivity to changes in fault states, robustness against other interferences, repeatability of measurements, signal attenuation or loss, accessibility, and impact on the surrounding environment.

Sensors: Permanent, semipermanent, or portable measuring instruments can be used. The connection of active sensors should comply with the power load and wiring specifications of the product. For situations where sensors cannot be installed, offline analysis can be conducted through data sampling. For example, installing cameras near key mechanical components can use image recognition technology for fault diagnosis and prognosis. The sensors themselves should ensure the necessary reliability and maintainability and ensure the safety of their installation work.

Data Collection: Data sources include maintenance records, environmental monitoring, and sensors installed on key components. The transmission of data collected by sensors should be efficient, accurate, and timely.

Data collection frequency: The inspection interval is determined based on the expected observability of faults and the rate of deterioration.

Fault Diagnosis: Fault diagnosis can involve feature extraction for data preprocessing and the use of diagnostic methods for further data analysis, mainly including the following aspects:

Feature extraction: The purpose of feature extraction is to provide input for fault diagnosis. Feature information strongly related to the operating state of the product is extracted to improve the accuracy of fault diagnosis. In the feature extraction process, attention should be paid to the representativeness of the features and their relevance to failure modes, ensuring that the extracted data retains the properties of the original data. Common feature extraction methods include but are not limited to methods based on time-frequency domain analysis.

Common fault diagnosis methods:

- Failure physics-based method: Model the product and its fault characteristics to determine the mechanism of failure.
- Data-driven method: Train algorithms using collected data to analyze fault patterns.
- Rule-of-thumb method: Evaluate and judge the mechanism of failure based on the experience of domain experts.

Note: The above three methods can be used in combination.

Fault Prognosis: Fault evolution trends can be projected and fault prognosis can be performed based on historical data and reasoning models, mainly including the following aspects:

Common fault prognosis methods:

- Physics-based prognosis methods, which predict based on the physical degradation mechanism of the product. For example, methods based on cumulative damage, methods based on warning devices, and methods based on specific models.
- Data model-based prognosis methods, which use the historical or current data of the object to establish a model that approximates the hidden mapping mechanism between the object data and faults under certain functional constraints. For example, methods based on statistical regression, methods based on similarity, and methods based on stochastic processes.
- Model-hybrid prognosis methods, which use mechanism models to determine the number of parameters or fault modes, and then improve the model's accuracy through data.

Alarms and warnings: Generally, they need to be based on failure set points. The failure set point is the final parameter value when the product or system fails. This value needs to be determined through standards, manufacturing instructions, historical statistical data, and experience, and the model should be updated in real time based on real-time data to improve the model's accuracy. The warning and alarm thresholds are generally less than the failure set point. The warning limit determines the advance time required for maintenance. When determining this warning value, parameters such as the confidence level of the forecast, future product demand, maintenance time, and trend prognosis need to be determined. Alarms should be set to remind relevant personnel for further handling.

Due to the universality of methods and the intersection of application scenarios, some methods of fault diagnosis and fault prognosis may overlap.

Maintenance Decision: Maintenance decisions are based on the results of fault diagnosis and prognosis to determine maintenance strategies, in order to improve maintenance efficiency and enhance the active safety assurance capabilities of the product. Maintenance strategies usually include corrective maintenance, preventive maintenance, or predictive maintenance.

5.2.2 Integration with Existing Maintenance Operations

5.2.2.1 Maintenance and Maintenance Strategies

The discipline linked to the questions of describing the failure behavior of a technical system and thus technical reliability is the field of maintenance. The maintenance of a product, component, assembly, or system ensures flawless function and thus has a direct influence on the reliability of the system. Basic activities of maintenance include maintenance, inspection, repair (colloquially: fixing), and the improvement of a technical system. Essentially, three different maintenance strategies are distinguished: breakdown maintenance, time-based maintenance, and condition-based maintenance (CBM). In breakdown maintenance, the system failure is taken as the occasion to carry out maintenance. On the other hand, in time-based maintenance, maintenance work is performed at defined intervals, usually related to a time interval. In CBM, operating parameters and conditions are monitored (continuously through operational data acquisition or through regular inspections), and maintenance work is planned and carried out depending on the system's condition. Condition-monitoring systems are often used for CBM.

5.2.2.2 Condition-based Maintenance (CBM)

CBM avoids fixed inspection and overhaul intervals and thus the periodic replacement of fully functional parts and assemblies. Furthermore, it can prevent a reduction in availability due to too frequent planned measures. By continuously or periodically measuring or observing certain dimensions and their changes on parts, assemblies, and components during operation, it is possible to determine, for example, the progress of wear during ongoing operation. These measurements or observations and the evaluation of the results are a special form of inspecting (condition monitoring). This makes it possible to reduce maintenance effort without compromising reliability and safety.

The purpose of CBM is to plan and carry out maintenance measures in a time-, quality-, and cost-optimized manner. With this maintenance strategy, intensive monitoring of important parts and equipment during ongoing operation is carried out (e.g., through automatic measuring devices). This allows for the prediction of when a failure would occur. This enables timely maintenance measures, such as renewal, to take place before the failure occurs.

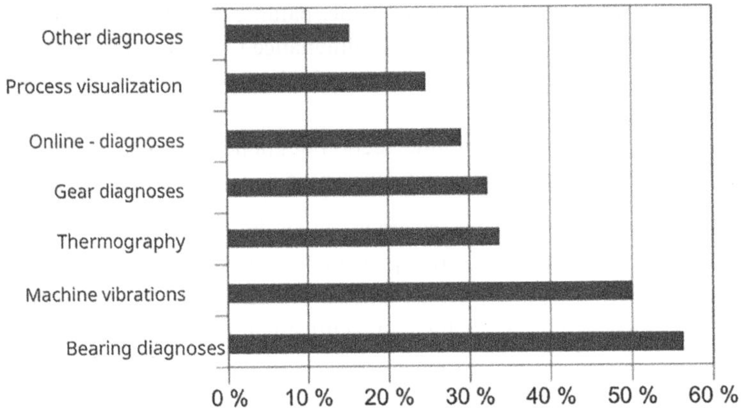

Figure 5.3 Relative Frequency of Today's Methods Used for Machine Condition Monitoring. *Source:* Bertsche and Dazer [3] / with permission of Springer Nature.

The application of CBM is possible for systems and components where operating conditions can be measured and monitored over time. Monitoring techniques for CBM include, among others:

- Thermal monitoring.
- Nondestructive material testing.
- Oil analysis.
- Vibration analysis.

Wear indicators on vehicle brakes measure the degree of lining wear and thus allow the prediction of the remaining service life before renewal is necessary. Figure 5.3 shows the relative frequency of today's methods used for machine condition monitoring. Bearing diagnostics and the examination of machine vibrations are therefore the most frequently used methods [2].

This approach requires a high effort in terms of data collection and evaluation, but it also reduces overall maintenance costs without compromising the safety and reliability of the equipment in any way.

Typical Condition Monitoring Program Flowchart: Particular techniques of condition monitoring are only introduced briefly and are covered in more detail in other international standards referenced within this standard. A typical condition monitoring program flowchart is shown in Figure 5.4. The sections of the flowchart are detailed in clauses 5–10.

— **ISO 17359:2002(E) Condition monitoring and diagnostics of machines—General guidelines.p3**

Predictive Maintenance Integrated Production Scheduling (PdM-IPS) Framework. Figure 5.5 presents the concept of the overall PdM-IPS approach with its modules. Information from the scheduler, production orders and from the shop floor serve as inputs to the approach. The planning module receives performance measures and the weights

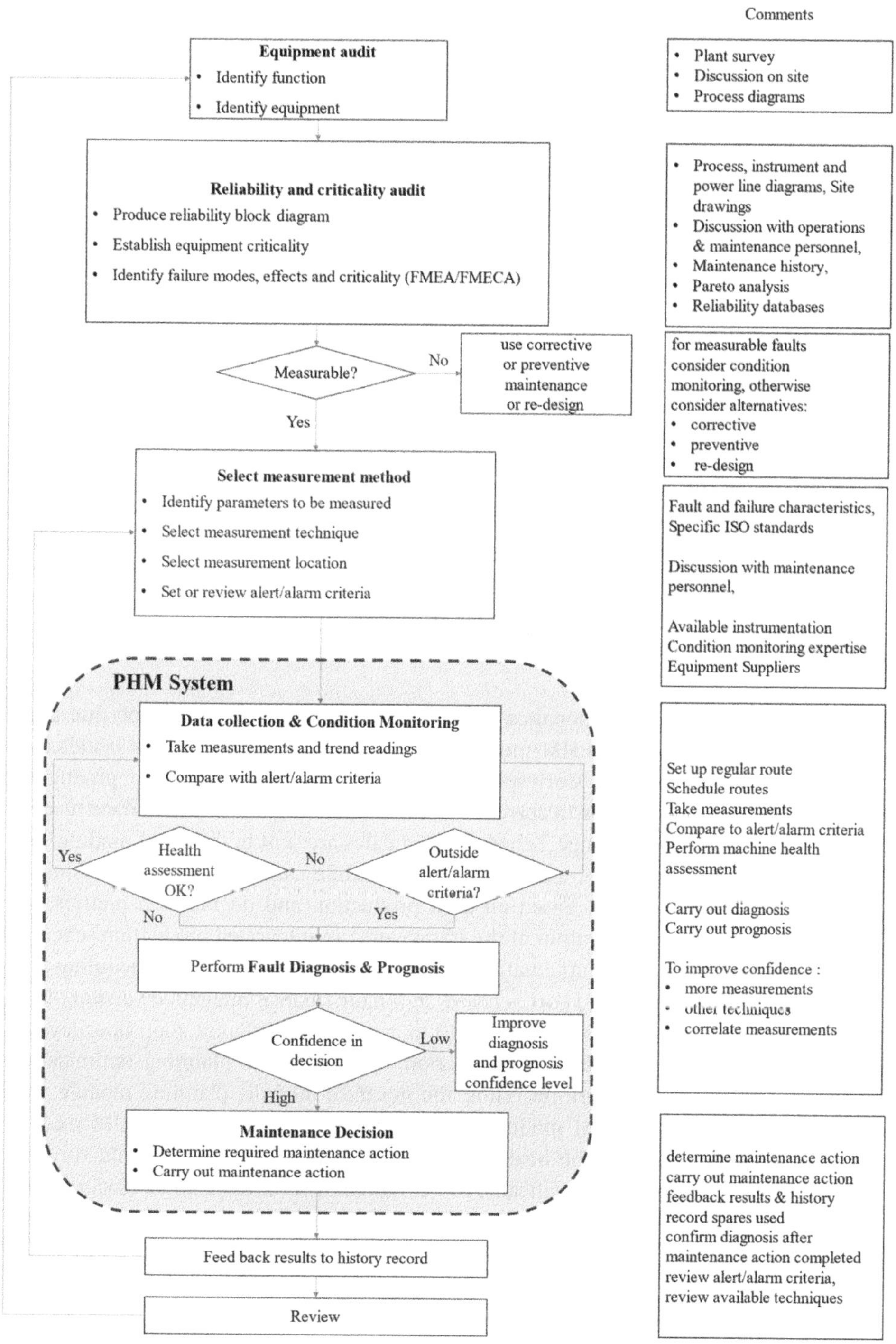

Figure 5.4 Typical Condition Monitoring Program. *Source:* Modified from Zhai et al. [4].

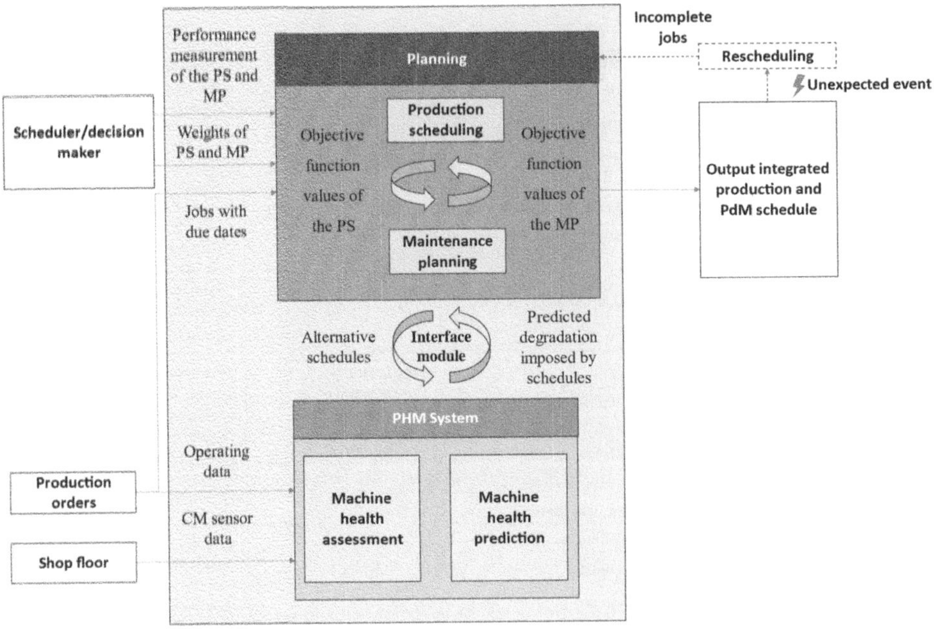

Figure 5.5 Overall PdM-IPS Framework (MP: Maintenance Plan, PS: Production Schedule). *Source:* Zhai et al. [4] / with permission of Springer Nature.

for both production and maintenance plans. In addition, it receives the job due dates from production orders. The PHM module receives CM data from sensors installed in machines on the shop floor. Corresponding operating data derived from production orders enable the PHM module to correlate CM and operating data. An interface module enables the information exchange. Schedule candidates are sent to the PHM module, and the predicted degradation imposed by this very schedule candidate will be transmitted back to the planning module. Based on both production and degradation metrics, the planning module derives the output of the framework: an integrated production schedule with maintenance actions. Mathematically, the integrated production scheduling and maintenance planning (IPSMP) part is based on a maintenance integrated version of the flexible job shop scheduling problem (FJSSP) to account for different shop floor layouts (i.e., MIFJSSP). As mentioned, this publication focuses on the planning optimization using a two-stage genetic algorithm. Thus, the interface module, planning module, and the interactions with the PHM module will be explained in detail. The PHM module is able to assess and predict the health condition of a machine after manufacturing a specific production schedule. Specifically, two conditional variational autoencoder neural networks are applied to derive operation-specific health indicator prediction of machines. The algorithms presented in this publication are able to work with any other PHM module that is capable of returning an operation or product-specific health indicator prediction.

— Predictive maintenance integrated production scheduling by applying deep generative prognostics models: approach, formulation, and solution.

5.2.3 Scalability and Adaptability in PHM Design

5.2.3.1 Activities of PHM and Reliability Over the Product Lifecycle

In Figure 5.6, the activities of PHM and reliability are depicted over the product lifecycle. In the context of reliability, the first step is the planning of necessary measures and objectives or requirements for reliability. In the second step, reliability is analyzed using qualitative and quantitative methods. One of the most important steps is to ensure the required service life through an appropriate test configuration. It is particularly important to pay attention to the number of test samples and the level of stress. If the products meet the required reliability, they can be released for customer use. In customer use, a variety of different application scenarios are possible, which can affect reliability. Failures are captured and further analyzed by the customer to identify possible weaknesses and to address them with appropriate measures. In addition, customer usage data can be fed back into development, enabling continuous product improvement.

In contrast, a PHM system is only applied during customer use. The extent to which the predicted remaining useful life (RUL) is used for operation optimization does not play a role in the application. Before application, the phases of requirement, planning, and system development are gone through. To determine requirements for PHM, a rough system topology is necessary. In this step, the components to be monitored and the necessary accuracy of the RUL forecast are defined. The accuracy of the forecast is based on possible measures in optimization and customer requirements. In PHM planning, the function and consequences of the failure mechanisms to be analyzed and the functional structure of the product are described. This is followed by the development of the PHM system. This primarily includes the formation and validation of a model for estimating and forecasting the

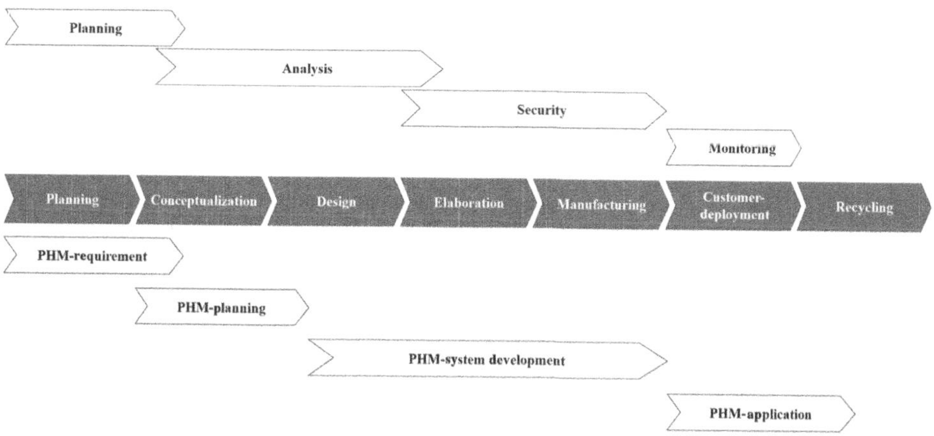

Figure 5.6 Assignment of Reliability and PHM Activities Over the Lifecycle of a Product. *Source:* Bertsche and Dazer [3] / with permission of Springer Nature.

damage condition. In the final step, the developed PHM solution is applied during operation in the field.

Activities within both the scope of reliability and PHM focus on identifying relevant failure mechanisms, determining critical loads, characterizing the mechanisms of failure types, and determining the usage behavior in operation. The identification of relevant failure mechanisms and the determination of critical loads are carried out for the entire population in both PHM and reliability.

As mentioned in Chapter 1 (Section 1.9, Prognostics and Health Management [PHM]), PHM and robust design serve distinct but complementary roles in the product lifecycle. Compared to PHM, which emphasizes online monitoring and dynamic updating as described in Chapter 5 (Section 5.2.3.1, Activities of Prognostics and Health Management [PHM] and Reliability Over the Product Lifecycle), robust engineering—from an engineering physics perspective—focuses on developing conceptual design robustness and further optimizing robustness in detailed design, including parameter design and tolerance design, during the early design stage.

5.2.3.2 Integration of Robust Design and PHM for Enhanced System Reliability

PHM and robust design are implemented at different stages of the product lifecycle. Robust engineering is predominantly applied during the system planning, concept design, and detailed design phases, whereas PHM is applied during the operational phase to monitor, diagnose, and predict system health. Robust design aims to enhance the fundamental function of a product or process, facilitating flexible designs and concurrent engineering. By focusing on desensitizing designs to the harmful side effects of noise factors (uncontrollable variations in the user environment), robust design improves system resilience.

Taguchi highlighted that variability arises due to active physics in design and environmental factors, which promote change over time [4]. Noise factors can be classified into five categories:

1) Piece-to-piece variation—e.g., variation in rubber thickness.
2) Change over time—e.g., material wear, force degradation, or dimensional changes over time.
3) Customer use variation—e.g., differences in open-hole wellbore size.
4) Environmental conditions—e.g., temperature fluctuations.
5) System interfaces and interactions—e.g., dimension variations of external elements and open-hole size.

The consequences of noise may result in degradation in quality (soft failure) or malfunction failure (hard failure). A product is said to be robust when it remains insensitive to variability, even if the sources of variability are not eliminated.

5.2.3.3 The Power of Integrating Robust Engineering and PHM

The integration of robust engineering and PHM establishes a structured engineering discipline that enhances system reliability by:

- Minimizing or eliminating potential "birth defects" in concept design.

- Strengthening product health conditions through robust optimization during detailed design (e.g., parameter design and tolerance design).
- Identifying critical parameters (both product and process) as health indicators during the robust design phase.

These identified health indicators serve as high-confidence inputs for PHM applications during the operational phase, including manufacturing, maintenance, and field use. By integrating robust engineering principles with PHM strategies, the reliability and health management framework gains a proactive foundation, enhancing confidence in conventional reliability engineering and PHM activities.

5.2.3.4 Assignment of Reliability and PHM Activities Over the Product Lifecycle

In Figure 5.7, the activities of PHM and reliability are depicted over the product lifecycle.

In the context of reliability, the first step is planning necessary reliability measures and defining key objectives/requirements. The second step involves qualitative and quantitative reliability analysis. One of the most critical steps is ensuring the required service life through appropriate test configurations, where attention to sample size and stress levels

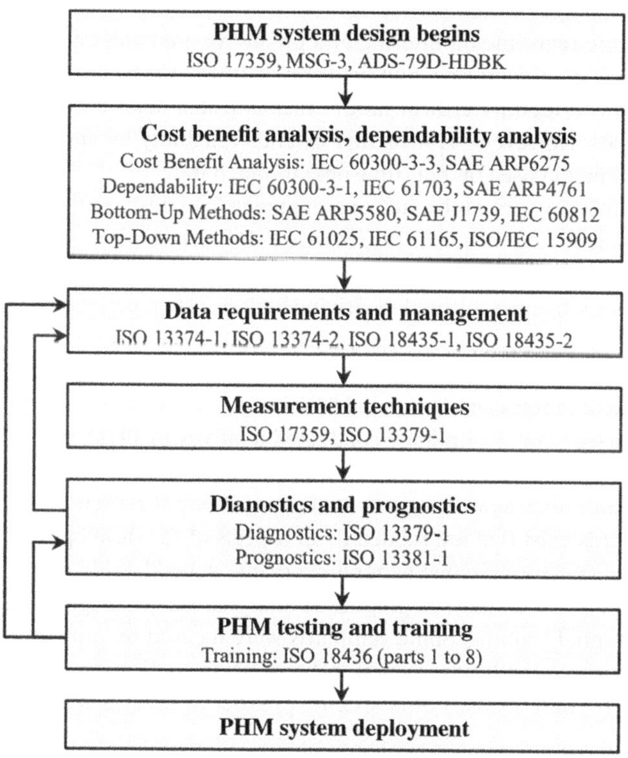

Figure 5.7 General PHM System Development Process and Associated Standards. *Source:* Vogl et al. [1] / The Prognostics and Health Management Society / CC by 3.0.

is crucial. Once a product meets reliability requirements, it is released for customer use. Failures occurring in real-world applications are captured and analyzed, allowing manufacturers to identify weaknesses and implement corrective actions. Additionally, customer usage data can be fed back into the development phase to drive continuous product improvement.

5.2.3.5 The Role of PHM in the Product Lifecycle

Unlike reliability methods applied throughout the product lifecycle, a PHM system is specifically deployed during the operational phase to optimize performance and predict failures. Before implementation, PHM planning requires defining system requirements, which involve:

- Establishing a system topology to determine critical components to be monitored.
- Defining the required accuracy of RUL predictions based on optimization measures and customer expectations.
- Analyzing failure mechanisms and describing functional structures that influence system performance.

Once planned, PHM system development focuses on model formation and validation for estimating and forecasting damage conditions. The final step involves deploying PHM solutions during operation, where real-time monitoring and predictive analytics enhance decision-making in maintenance, manufacturing, and field applications.

Both reliability and PHM aim to identify critical failure mechanisms, determine load factors, and analyze failure types. However, PHM further extends reliability by enabling proactive system health management based on real-time operational data.

By implementing robust engineering principles upfront, engineers can design products that inherently resist variability, reducing the need for excessive PHM interventions during operation. This synergistic approach not only improves reliability and robustness but also enhances predictive maintenance strategies, ultimately driving higher system performance and longevity.

5.2.3.6 PHM System Development Process and Associated Standards

The National Institute of Standards and Technology conducted a survey of PHM-related standards to determine the industries and needs addressed by such standards, the extent of these standards, and any similarities as well as potential gaps among the documents. This effort revealed that standards exist that are related to all aspects of the development of prognostics and health management systems: general overview, dependability analysis, measurement techniques, diagnostic analysis, prognostic analysis, data management, performance metrics, and personnel training. Some standards were focused on providing guidance for specific applications, yet still broad enough for general application across industries. Other standards were more focused on a specific product or process within a target industry.

Table 5.2 PHM-Related Standard Organization.

Organization	Committee/ Subcommittee	Overview	Cost and Dependability analyses	Measurement techniques	Diagnostics and Prognostics	Data management	Training	Applications
ATA	MSG	X						X
IEC	56	X	X					
IEEE	RS	X						
ISO	TC 108/SC 2			X				
ISO	TC 108/SC 5	X		X	X	X	X	
ISO	TC 184/SC 4					X		
ISO	TC 184/SC 5				X	X		
ISO/IEC	JTC 1/SC 7		X					
MIMOSA	——					X		
SAE International	AQPIC		X					
SAE International	E-32	X			X			X
SAE International	G-11r	X	X					
SAE International	HM-1	X	X	X	X			X
US Army	Aviation Engineering	X		X	X			X

Source: Vogl et al. [1] / The Prognostics and Health Management Society / CC by 3.0.

Table 5.3 PHM-Related Standard Under Development.

Organization	Committee/ Subcommittee	Standard	1st Edition / Revision?	Overview	Dependability Analysis	Diagnostics and Prognostics	Data Management
SAE Int.	G-11r	ARP6204	1st Edition	X			
SAE Int.	HM-1	ARP6268	1st Edition	X			
SAE Int.	HM-1	ARP6407	1st Edition	X			
SAE Int.	HM-1	ARP6883	1st Edition	X			
IEEE	RS	P1856	1st Edition	X			
ISO/IEC	JTC 1/SC 7	ISO/IEC 15909-2	1st Edition		X		
ISO	TC 108/SC 5	ISO 13379-2	1st Edition			X	
ISO	TC 108/SC 5	ISO 13381-1	Revision			X	
ISO	TC 108/SC 5	ISO 18129	1st Edition			X	
ISO	TC 184/SC 5	ISO 22400-1	1st Edition			X	
ISO	TC 184/SC 5	ISO 22400-2	1st Edition			X	
ISO	TC 184/SC 5	ISO 18435-3	1st Edition				X
SAE Int.	HM-1	ARP6290	1st Edition				X

Source: Vogl et al. [1]/The Prognostics and Health Management Society/CC by 3.0.

References

1 Vogl, G.W., Weiss, B.A., and Donmez, M.A. (2014). Standards for Prognostics and Health Management (PHM) Techniques within Manufacturing Operations. In: *Proceedings of the Annual Conference of the Prognostics and Health Management Society 2014*, vol. 6. Fort Worth, TX, USA.

2 International Organization for Standardization (2002). *ISO 17359:2002(E), Condition Monitoring and Diagnostics of Machines—General Guidelines*. ISO: Geneva, Switzerland.

3 Bertsche, B. and Dazer, M. (2022). *Zuverlässigkeit im Fahrzeug- und Maschinenbau: Ermittlung von Bauteil- und System-Zuverlässigkeiten*, 4the. Springer Vieweg: Berlin, Heidelberg, Germany.

4 Zhai, S., Kandemir, M.G., and Reinhart, G. (2022). Predictive maintenance integrated production scheduling by applying deep generative prognostics models: approach, formulation and solution. *Production Engineering* 16 (1): 65–88.

6

Characterizing Failure Signatures

CHAPTER MENU

6.1 Characterizing Failure Signatures

6.1.1 Identifying Degradation Patterns

Degradation patterns refer to the systematic changes in the performance of a system or component over time due to wear, aging, or other factors. These patterns can be identified through various methods and are crucial for predicting failures and maintaining system reliability.

Examples of Degradation Patterns

- **Linear Degradation:** In certain systems, degradation occurs at a constant rate, such as a mechanical bearing gradually losing efficiency as it ages.
- **Exponential Degradation:** Components like batteries or capacitors may degrade exponentially, where performance drops sharply as they approach the end of their life.
- **Stepwise Degradation:** In some systems, degradation might not be continuous but occurs in distinct steps, often triggered by external stressors like temperature spikes or sudden impacts.
- **Cyclic Degradation:** Mechanical systems like engines or pumps often exhibit degradation that worsens during each cycle of operation, such as fatigue cracking or wear in rotating parts. Typical physics-of-failure (PoF) models include the Arrhenius model, the Eyring model, and the Coffin–Manson model.

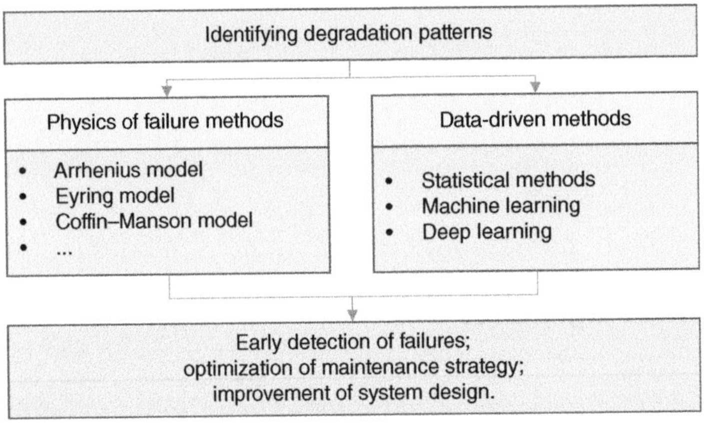

Figure 6.1 The Role and Approaches of Identifying Degradation Patterns.

Identifying degradation patterns plays a critical role in PHM systems, serving as the linchpin between raw operational data and actionable maintenance intelligence. This capability extends beyond simple anomaly detection—it systematically deciphers the temporal evolution of asset health, enabling organizations to:

- **Early Detection of Failures:** By identifying degradation early, maintenance can be scheduled proactively, preventing unexpected downtime.
- **Optimization of Maintenance strategy:** Recognizing degradation trends allows maintenance schedules to be optimized, reducing costs associated with both over-maintenance and under-maintenance.
- **Improvement of System Design:** Understanding degradation patterns helps engineers design more robust systems with longer lifespans, improving overall system reliability.

As shown in Figure 6.1, to identify degradation patterns, there are two primary categories of methods: PoF methods and data-driven methods.

6.1.1.1 Synergistic Integration: Robust Design, Physics of Failure, and Degradation Pattern Identification

Developing robust, resilient, and long-lasting systems requires the integration of three foundational engineering approaches:

Robust Design: Mitigating Sensitivity to Variation

Robust design ensures systems perform consistently despite uncontrollable factors like environmental variation and manufacturing tolerance.

- **Tools:** DOE, Taguchi Methods, Tolerance Analysis
- **Purpose:** Minimize performance sensitivity to variation
- **Contribution:** Identifies critical parameters and guides early reliability-oriented design

Physics of Failure (PoF): Modeling Physical Failure Mechanisms

PoF uses scientific principles to model how materials and components fail under stress, aging, or environmental loads.

- **Mechanisms:** Fatigue, corrosion, thermal cycling, creep, etc.
- **Tools:** Finite element analysis (FEA), stress–strain modeling.
- **Contribution:** Links environmental and mechanical stressors to potential failure modes.

Degradation Pattern Identification: Tracking Health in Real Time

This method captures and interprets performance decline over time, offering empirical support for PHM.

- **Applications:** Predicting Remaining Useful Life (RUL), proactive maintenance
- Methods:
  - Sensor-based data collection (e.g., vibration, temperature).
  - Trend analysis and statistical modeling.
  - Machine learning and AI prognostics.

Integrated Synergy: How They Work Together

Element	Robust Design	Physics of Failure	Degradation Analysis
When Used	Early in design	Design & Development	Post-deployment
Primary Focus	Reduce variation sensitivity	Model physical failure modes	Detect and predict degradation
Output	Design tolerances and margins	Stress/failure models	Health indicators and trends
Interconnection	Guided by PoF insights	Validates robust assumptions	Confirms predictions with data
Joint Benefit	Proactively prevent failure	Understand how failure evolves	Respond to and manage risks

Practical Example: Electronics in Harsh Environments

- **Robust Design:** Ensures circuit boards function reliably across wide environmental conditions.
- **PoF:** Models solder joint fatigue due to thermal cycling.
- **Degradation Monitoring:** Tracks resistance shifts via sensors to forecast failure.

This triad forms a **closed-loop reliability framework**: design robustly, model intelligently, and monitor dynamically.

The Strategic Role of Degradation Pattern Identification in PHM

Identifying degradation patterns is central to prognostic health management (PHM) systems. It goes beyond anomaly detection, helping organizations:

- **Detect Failures Early:** Proactively schedule maintenance to prevent downtime.
- **Optimize Maintenance:** Avoid over- or under-maintenance with accurate trend insights.
- **Improve Design:** Feed operational data back into design for better robustness and longevity.

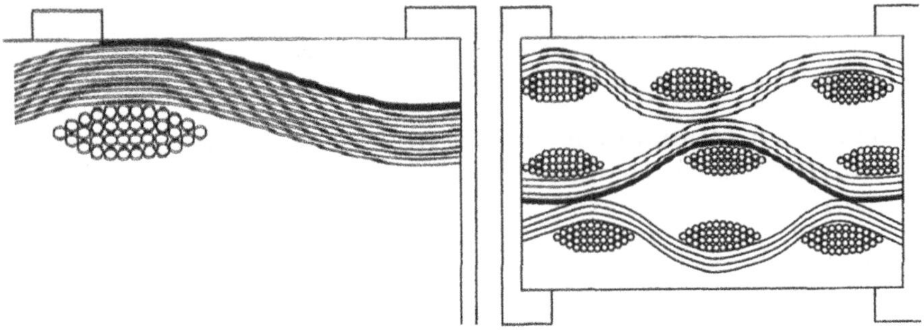

Figure 6.2 Formation of Conductive Filament [1].

Methods of Degradation Pattern Identification

Degradation analysis typically involves two complementary method categories:

- **Physics of Failure-based Methods:** Analytical, model-based approaches grounded in material science.
- **Data-driven Methods:** Statistical, AI, and sensor-based monitoring techniques for trend analysis.

Together, these approaches reinforce the PHM framework and enable a shift from reactive maintenance to predictive, intelligent system management.

PoF Methods PoF methods rely on understanding the physical principles behind degradation processes. These methods often use analytical models of system behavior under stress, such as fatigue, corrosion, or wear. The primary advantage of these methods is that they can offer a detailed, physically grounded understanding of failure progression. However, they can be computationally expensive and may require precise knowledge of the system's physical properties. Typical PoF models include:

The process to establish a PoF model typically involves:

- Model Development: Creating a mathematical model that describes the degradation process.
- Parameter Estimation: Using experimental data to estimate model parameters.
- Simulation and Validation: Running simulations to predict degradation and validating the model against observed data.

Conductive Filament Formation [1]

Model Development: The case study focuses on conductive filament formation, a failure mechanism commonly observed in laminated substrates used in printed circuit boards (PCBs). Formation of conductive filament is illustrated in Figure 6.2. A mathematical model was developed to describe the degradation process caused by electrochemical migration of metalions. The model characterizes the time to failure as a function of key parameters such as relative humidity, voltage, and the distance between conductors. The model is expressed as:

$$t_f = \frac{a \cdot L^n}{(M - M_0) \cdot V^m}$$

where:

- t_f is the time to failure due to conductive-filament formation,
- a, n, m are acceleration factors,
- L is the spacing between conductors,
- V is the voltage bias applied between conductors,
- M is the percentage moisture absorbed by the substrate,
- M_0 is a threshold moisture content.

Parameter Estimation: Experimental data were collected from 52 substrates tested over 1500 hours. The substrates were made of three commonly used materials: FR-4, bismaleimide triazine (BT), and cyanate ester (CE). The parameters studied included temperature, relative humidity, operating voltage, conductor spacing, metallization layer position, and the presence or absence of coatings such as solder mask, solder plate, and post coat. The experiments involved exposing the substrates to controlled temperature–humidity environments and applying a voltage bias across pairs of conducting lines. The time to failure was recorded when the insulation resistance dropped significantly due to conductive filament formation.

Simulation and Validation: The developed model was used to simulate the time to failure under various environmental conditions and design parameters. The simulations predicted the degradation process and helped identify safe conductor spacing for different materials and environmental conditions. The model was validated against observed data from the experiments, showing good agreement. Additionally, the model was used to guide design improvements, such as selecting alternative materials or implementing environmental controls when narrow spacing was required. Accelerated test environments were also evaluated to ensure the model's predictions aligned with real-world usage conditions.

Data-Driven Methods

Data-driven methods use historical and real-time data to identify degradation patterns without requiring detailed physical models. These methods can broadly be classified into statistical methods, machine learning methods, and deep learning methods.

- **Statistical Methods:** These methods, such as regression analysis, are straightforward and rely on statistical relationships between variables. They are easy to implement and interpret but often fail to capture complex patterns that may emerge in nonlinear or multi-variable systems.
- **Machine Learning Methods:** These include techniques like decision trees, support vector machines, and random forests. Machine learning methods are capable of identifying complex, nonlinear degradation patterns. They require large datasets for training but can be highly accurate once trained. However, these methods can be prone to overfitting and may not generalize well to new, unseen data.
- **Deep Learning Methods:** Deep learning methods involve neural networks with multiple layers to learn patterns from data. It can automatically extract features from large datasets. Thus deep learning methods are powerful for identifying complex degradation patterns, especially in high-dimensional data such as images or time-series

Figure 6.3 Common Process of Data-driven Method.

data. However, deep learning methods often require significant computational resources and large amounts of data for training and be difficult to interpret.

As shown in Figure 6.3, the process of data-driven methods to identify degradation patterns typically follows these steps:

1) **Data Collection:** Gather performance data from sensors, maintenance logs, and other relevant sources.
2) **Data Preprocessing:** Clean the data by removing noise and outliers, ensuring it's suitable for analysis.
3) **Feature Extraction:** Identify key features that indicate degradation, such as changes in vibration levels, temperature, or pressure.
4) **Pattern Identifying:** Use the aforementioned techniques to identify patterns in the data that correlate with degradation.
5) **Validation:** Cross-validate identified patterns with failure data or simulated data to ensure their accuracy.

Battery Capacity Trajectory Prediction [2]

Data Collection: The case study focuses on monitoring the health of lithium-ion batteries used in electric vehicles. Performance data is collected from 121 commercial 18650-type batteries cycled under various temperatures and current rates. As shown in Table 6.1, the dataset includes two types of batteries: nickel cobalt aluminum oxide (NCA) and nickel cobalt manganese oxide (NCM), each with a rated capacity of 3.5 Ah. The batteries are cycled using a constant current-constant voltage (CC-CV) charging strategy and a constant current discharging strategy. Data is collected at three ambient temperatures: 25°C, 35°C, and 45°C, with different charge current rates applied. The dataset comprises information on battery capacity and health indicators (HIs) extracted from charging and resting phases.

Data Preprocessing: The collected data is preprocessed to ensure it is suitable for analysis. This involves normalizing the input data (health indicators) using the mean and standard deviation of the training data. This step ensures that the data is consistent and ready for use in the machine learning models.

Feature Extraction: The study introduces two key health indicators: Qmean and RVmean. Qmean represents the average charge increment in the 10 minutes before charging and reflects Coulombic efficiency during charging. RVmean is the average voltage in the 10 minutes after charging and indicates voltage recovery during rest. Both are strongly linked to battery degradation. Qmean corresponds to lithium loss (LLI), and RVmean to electrochemical interface degradation (SEI film growth). They show stable trends linked to capacity changes over the battery's life. Additionally, the study explored other features like Qstd and RVstd and found RVmean has the strongest linear relationship with capacity, making it the preferred input for subsequent analysis.

Table 6.1 Battery Information and Experimental Setting [2].

Battery types	NCA 18650	NCM 18650
Nominal capacity (Ah)	3.5	3.5
Cutoff voltage (V)	2.65–4.2	2.5–4.2
Temperature (°C)	25 / 35 / 45	25 / 35 / 45
Charge & discharge policies	0.25C / 0.5C / 1C CC–CV & 1C CC	0.5C CC–CV & 1C CC
Number of cells	66	55

Pattern Identifying: To enhance model accuracy in the early degradation phase, the K-means algorithm was applied for unsupervised clustering of battery degradation patterns into fast, medium, and slow decline groups. Initial clustering using the first 40 cycles' capacity-cycle data was unstable, particularly for NCA samples. However, using only the RVmean sequence from the first 40 cycles as input achieved precise degradation pattern recognition, with results 93% similar to those from full lifecycle data. This method outperformed clustering using Qmean or Qmean+RVmean, which had 30% and 3% lower accuracy, respectively. The final clustering results provided an improved data grouping basis for modeling and supported early degradation trend identification. Based on these clustering results, the study developed an LSTM-based multi-task learning (MTL) model for health indicator extrapolation and an LSTM-based model with a probabilistic neural network (PNN) for capacity trajectory prediction.

Validation: To assess the model's accuracy, the study compared the model's extrapolated health indicators (HIs) with actual datasets, which is shown in Figure 6.4. The MTL model predicted Qmean and RVmean simultaneously, reducing Qmean extrapolation errors by 25%–46% and keeping RVmean errors below 0.1% compared to single-task learning models. The LSTM–PNN model used extrapolated HIs for capacity trajectory prediction, offering high precision and uncertainty assessment via a 95% confidence interval. It showed excellent prediction across three aging stages (10%, 30%, 50% of observable data), with average MAPE for capacity prediction below 2.59% and 1.15% for NCA and NCM cells, respectively, and average absolute EOL prediction errors within 54 and 35 cycles. This outperformed common models like GPR, RFR, and SVR, while having fewer parameters and lower computational costs, making it feasible for real-world deployment.

6.1.2 Signature Analysis for Different System Components

In robust system design, especially during the early design phase, it is crucial to optimize designs in the presence of various sources of noise and variability—including piece-to-piece variation, environmental aging, external temperature and humidity fluctuations, customer usage diversity, and complex system interfaces and structural interactions. Robust design

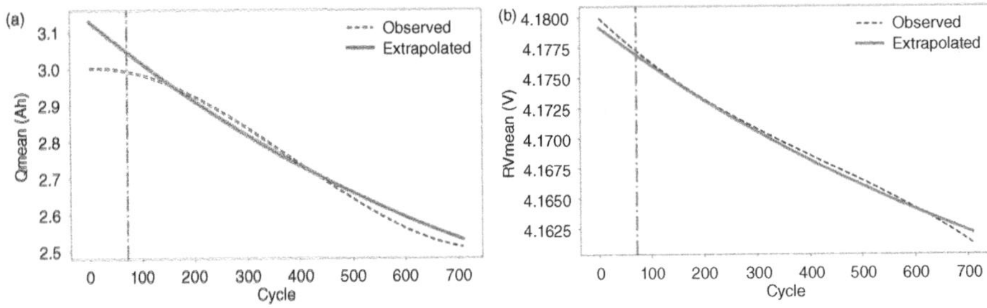

Figure 6.4 Results of HIs Extrapolation of: (a) Qmean and (b) RVmean of an NCA Cell in the Early Aging Stage Using MTL [2].

plays a pivotal role here by establishing the foundation for resilience before operational degradation occurs.

Building on this robust foundation, **signature analysis** helps identify and track degradation trends across different system components. However, each component behaves differently, requiring specific analysis techniques tailored to its operational context and failure modes.

Why Different Components Require Different Analysis Methods

1) Unique Failure Modes
   - Mechanical components may degrade due to friction, fatigue, or wear, while electronic components might suffer from thermal fatigue, voltage spikes, or corrosion.
   - Understanding these differences ensures the right analytical model is used for each component type, strengthening the alignment between design intent and real-world behavior.
2) Variety of Data Types
   - Mechanical systems often produce vibration or pressure data, while electronic systems generate current–voltage profiles, thermal maps, or signal noise data.
   - Signature analysis methods must be designed to extract relevant features from the specific type of data generated.
3) Operational Conditions
   - Components in noisy, high-temperature environments degrade differently than those in controlled conditions.
   - Robust design identifies critical parameters early on, but ongoing signature analysis adapts maintenance strategies to real-world conditions.
4) Complexity of Systems
   - In integrated systems, component interactions introduce emergent behaviors. A failure in one component may trigger unexpected effects in others.
   - Advanced data analytics and system-level modeling are essential to distinguish root causes from secondary symptoms.
5) Maintenance Strategies
   - Signature analysis methods must support condition-based or predictive maintenance.
   - For example, frequently serviced components may use high-sensitivity anomaly detection, while long-lifecycle components benefit from trend-based degradation monitoring.

Examples of Signature Analysis Techniques:

Component Type	Common Signature Data	Analysis Method
Bearings	Vibration signals	FFT, envelope detection, spectral kurtosis
Batteries	Voltage, temperature	Capacity fade models, EIS, SoH estimation
PCBs (Printed Circuits)	Temperature, resistance	Thermal imaging, impedance spectroscopy
Motors and Pumps	Acoustic and vibration data	Time–frequency analysis, wavelet transform
HVAC systems	Flow rate, pressure, temp	Regression models, decision trees

Incorporating robust design at the early design phase reduces the impact of variability from the start, setting the stage for successful PHM and degradation monitoring.

However, maintaining system reliability over time requires tailoring signature analysis methods to the distinct characteristics of each component. Together, robust design and component-specific signature analysis form a **closed-loop reliability strategy**—resilient by design, adaptive in operation.

(This paragraph has been modified to emphasize the closed-loop reliability strategy, as highlighted in blue above.) Different system components often require distinct analysis methods due to their unique operational characteristics, failure modes, and the type of data they generate. The choice of analysis method is crucial for accurately predicting failures and maintaining system reliability. Below is a detailed exploration of why different components requiring different analysis methods, along with examples of common signature analysis methods for various system components.

Different system components often require distinct analysis methods due to their unique operational characteristics, failure modes, and the type of data they generate. The choice of analysis method is crucial for accurately predicting failures and maintaining system reliability. Below is a detailed exploration of why different components require different analysis methods, along with examples of common signature analysis methods for various system components.

Why Different Components Require Different Analysis Methods

1) Unique Failure Modes:
   - Each component may fail due to different reasons. For example, mechanical components might fail due to wear and tear, while electronic components could fail due to thermal stress or electrical overstress.
   - Understanding these unique failure modes is essential for selecting the appropriate analysis method that can accurately predict and diagnose potential failures.
2) Variety of Data Types:
   - Different components generate different types of data (e.g., vibration data from mechanical systems, thermal images from electronic systems).

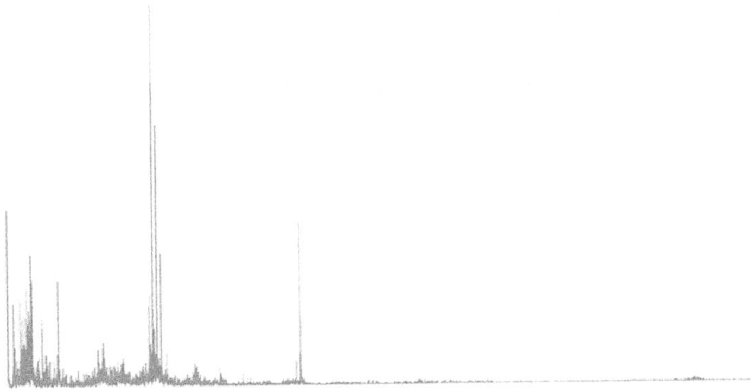

Figure 6.5 The FFT Spectrum Image of a Normal Signal [3].

- The analysis method must be capable of handling the specific type of data generated by the component to extract meaningful failure signatures.

3) Operational Conditions:
   - Components operate under various conditions that can affect their degradation and failure patterns. For instance, a component in a high-temperature environment may degrade differently from one in a low-temperature environment.
   - Tailored analysis methods are necessary to account for these varying operational conditions.
4) Complexity of Systems:
   - Modern systems are increasingly complex, with components interacting in intricate ways. A method that works for a simple system may not be effective for a complex, interconnected system.
   - Advanced analysis methods are required to capture the interactions and dependencies between components in complex systems.
5) Maintenance Strategies:
   - Different components may have different maintenance requirements and schedules. The analysis method should align with the maintenance strategy to provide actionable insights.
   - For example, a component that requires frequent maintenance may benefit from a more sensitive analysis method that can detect early signs of degradation.

6.1.3 Signature Analysis Methods for Various System Components

Mechanical Components

- Vibration Analysis:
  - o Vibration analysis is a common method for monitoring the health of rotating machinery. It involves analyzing the vibration signals generated by the machinery to detect anomalies indicative of degradation or failure.
  - o For example, changes in the frequency spectrum of vibration signals can indicate bearing wear or misalignment. For illustration, Figure 6.5 and Figure 6.6 show the FFT Spectrum of normal and abnormal signal respectively.

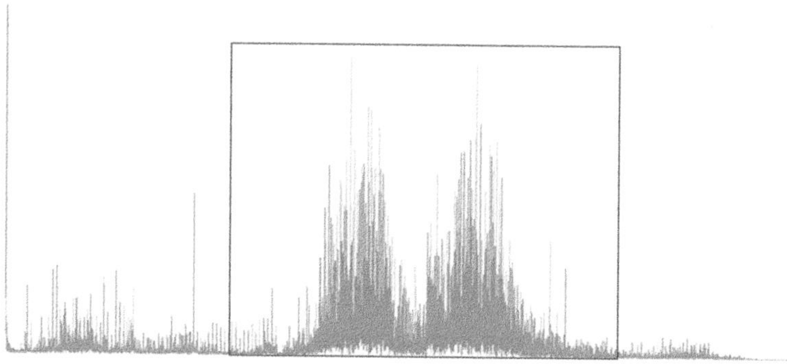

Figure 6.6 The FFT Spectrum Image of an Inner-race Fault Signal [3].

- Acoustic Emission Testing:
  - This method detects the acoustic signals generated by micro-cracks or other internal defects in materials. It is particularly useful for monitoring structural components under stress.
  - Acoustic emission testing can provide early warnings of potential failures, allowing for timely maintenance and repair.
  - For instance, acoustic emission (AE) can be used to monitor the microcrack and fracture evolution of coal and rock [4]. Figure 6.7 illustrates how acoustic emission (AE) energy evolves during the uniaxial compression of coal and sandstone specimens. The figure demonstrates that, as stress is applied, AE energy remains minimal during the initial compaction and elastic deformation phases, indicating the closure of existing microcracks. However, as loading continues and microcracks initiate and propagate, there is a significant increase in AE energy release, particularly near the peak stress. This surge in AE activity corresponds to the rapid development and coalescence of microcracks leading to macroscopic failure. The cumulative AE energy curves further reflect this progression, showing a marked increase as damage accumulates. These observations confirm that AE monitoring effectively detects acoustic signals generated by micro-cracks or internal defects, providing valuable insights into the structural integrity of materials under stress.

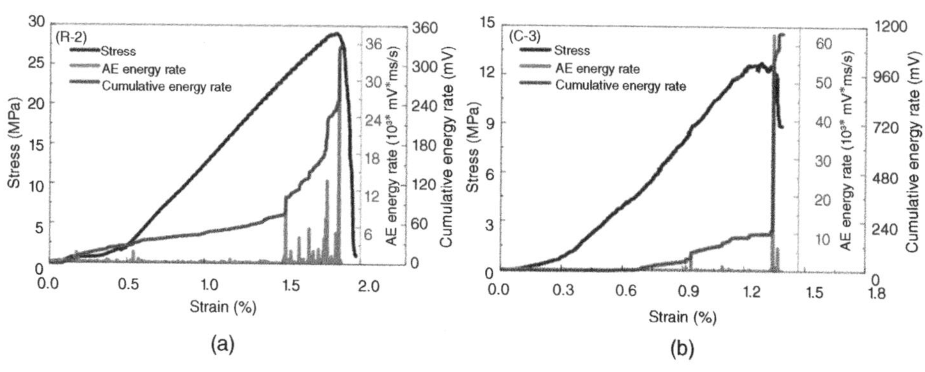

Figure 6.7 Energy Evolution of the Coal and Rock Failure Processes [4].

Electronic Components

- Thermal Imaging:
  - Thermal imaging uses infrared cameras to detect hotspots in electronic devices. Abnormal temperature distributions can indicate potential degradation, such as overheating of components or poor thermal management.
  - For example, a sudden increase in temperature in a specific area of a circuit board can suggest a failing component. Figure 6.8 (b) illustrates a fault that results in a heat point while the normal circuit has no such point in Figure 6.8 (a).
- Electrical Parameter Monitoring:
  - This method involves continuously monitoring parameters like voltage, current, and resistance. Anomalies in these parameters can indicate degradation or impending failure.
  - For instance, we can assessing the condition of high-voltage circuit breaker (CB) contacts by analyzing dynamic resistance measurements (DRM) [6]. The authors of [6] extracted five diagnostic parameters from DRM curves, such as the maximum and average resistance values of main and arcing contacts, as well as the overlapping time, which is shown in Figure 6.9. These parameters were then utilized to compute a health index, categorizing the CB's condition into three states: healthy, caution, and risky.

Battery Systems

- Electrochemical Impedance Spectroscopy (EIS):
  - EIS measures the impedance of a battery over a range of frequencies. Changes in impedance spectra can provide insights into the degradation mechanisms, such as electrolyte degradation or electrode material changes.
  - For example, Figure 6.10 illustrates the calendar aging behaviors and the corresponding impedance evolution [7]. (1) Temperature: Variation in the electrochemical impedance spectroscopy (EIS) of calendar-aged cells over storage time, accompanied by changes in the microstructure and X-ray diffraction (XRD) patterns of the cathode material. (2) State of Charge (SOC): Charge and discharge profiles of calendar-aged cells, along with Nyquist plots for disassembled anodes and cathodes, and corresponding simulation results. (3) Cathode: Changes in capacity and internal resistance observed in various batteries stored under different SOC and temperature conditions. (4) Further Cycling: Discharge curves and EIS fitting results comparing fresh cells and calendar-aged cells before and after additional cycling.

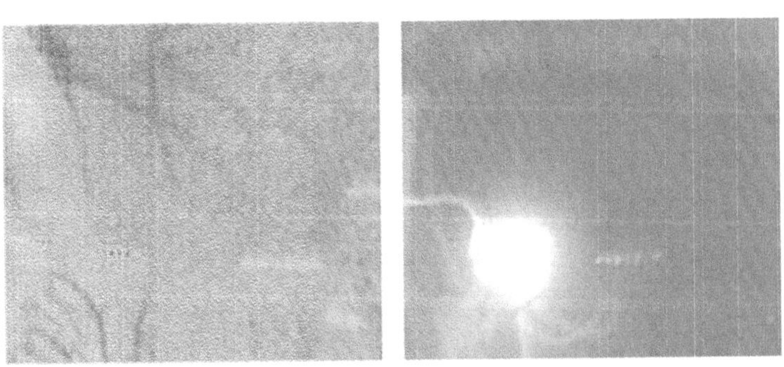

(a) (b)

Figure 6.8 (a) Normal Integrated Circuits. (b) Integrated Circuits with One Fault [5].

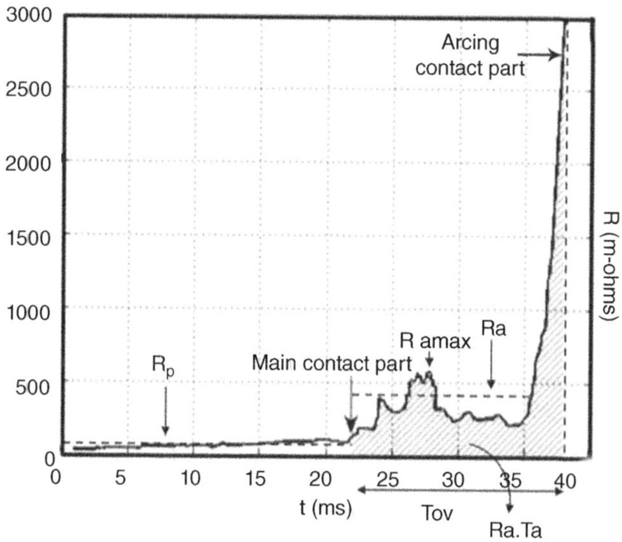

Figure 6.9 Typical DRM Curve with Suggested Diagnosis Parameters [6].

- Capacity Fade Analysis:
  - Monitoring the capacity of a battery over time can reveal degradation patterns. A significant drop in capacity indicates aging or degradation.
  - For instance, a battery that shows a rapid decrease in capacity over a short period may be experiencing accelerated aging due to operational conditions. As illustrated in Figure 6.11, the capacity decreases faster in higher temperature.

The choice of failure signature analysis method is critical for the effective monitoring and maintenance of different system components. Each component type presents unique challenges and requires tailored approaches to accurately predict and diagnose failures. By selecting the appropriate analysis method based on the component's characteristics and operational conditions, organizations can enhance their ability to maintain system reliability and reduce the risk of unexpected failures. The examples provided illustrate the diversity of methods available and the importance of matching the analysis technique to the specific needs of each component type.

6.1.4 The Role of Signatures in Failure Prediction

Failure signatures serve as the foundational inputs for predictive models in PHM systems. Their relevance lies in their ability to provide early warning indicators of impending failures, enabling proactive maintenance strategies and minimizing unplanned downtimes. The predictive power of a failure signature depends on its sensitivity, stability over time, and correlation with specific failure modes. In prediction of failure, the core functions of failure signatures can be summarized as:

- **Early Detection of Anomalies:** By continuously tracking operational parameters and comparing them to known failure signatures, anomalies can be detected well before catastrophic events occur.

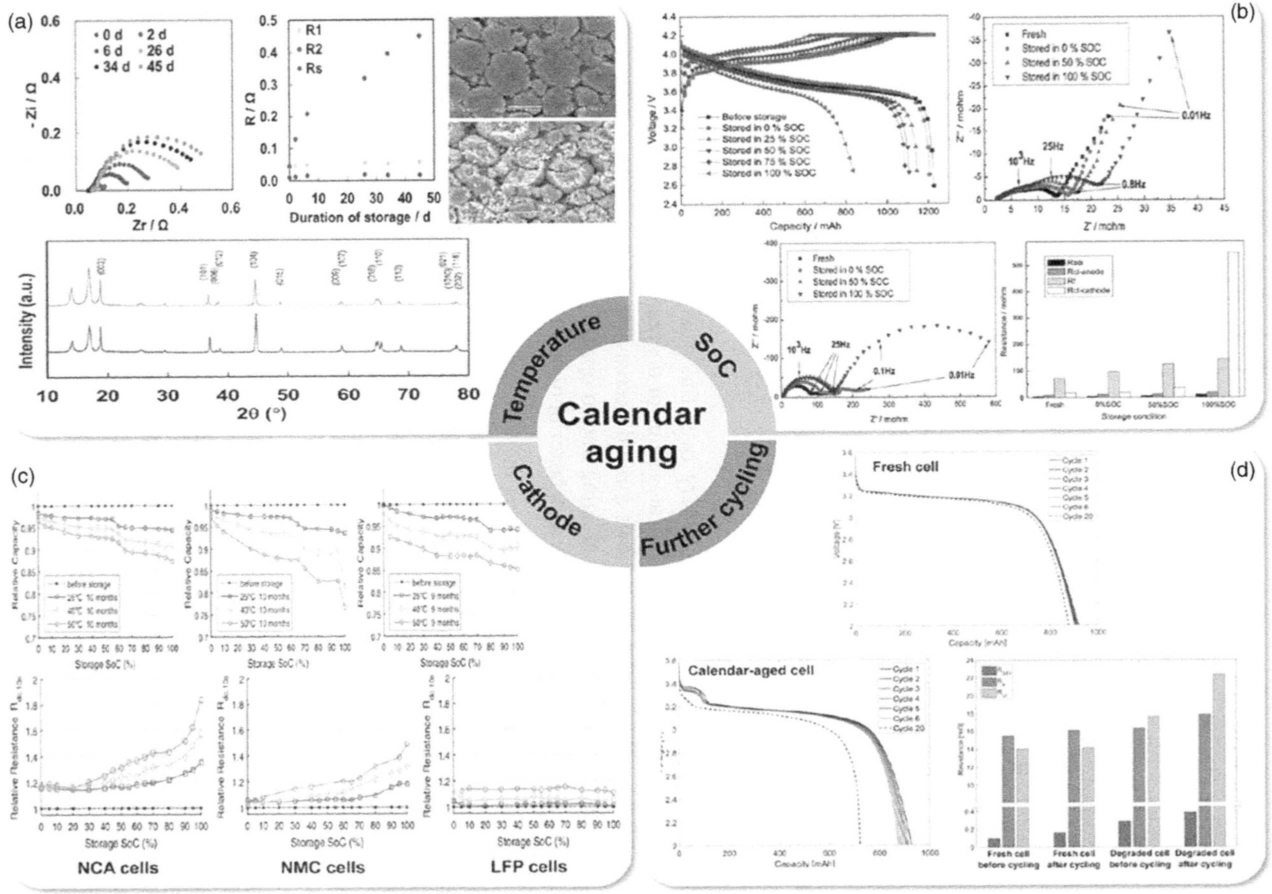

Figure 6.10 Calendar Aging Conditions [7].

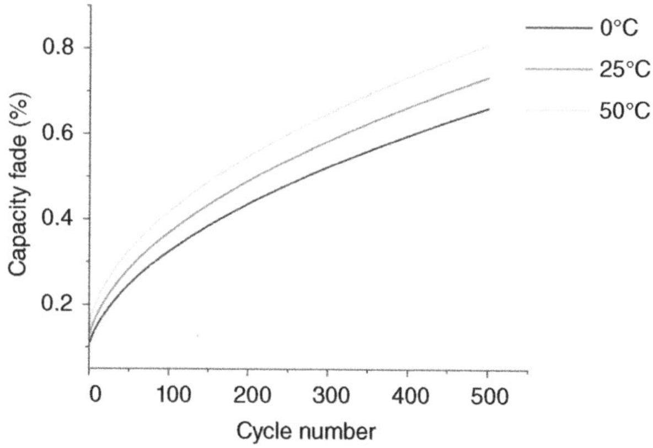

Figure 6.11 Capacity Fade with Cycle Number Increase [8].

- **Model Training and Validation:** Signatures are used to train machine learning and statistical models to recognize failure patterns. These models, once validated, can generalize from historical data to real-time prediction.
- **Remaining Useful Life (RUL) Estimation:** Accurate failure signatures allow for dynamic estimation of a component's RUL. This helps in optimizing maintenance intervals and spare part inventory.
- **Decision Support for Maintenance Planning:** By correlating signatures with failure risk levels, maintenance teams can prioritize actions based on criticality and risk, improving resource allocation.

To fully realize these benefits, several key considerations must be addressed when using failure signatures for prediction.

- **Signature Quality:** The reliability of predictions depends on the quality of the signatures, which in turn depends on data accuracy, relevance, and sampling frequency.
- **Noise and Uncertainty Management:** Effective preprocessing and feature extraction are critical to filter out irrelevant data and noise that could skew predictions.
- **Adaptability of Models:** Since system conditions and usage patterns may change, predictive models must be continuously updated with new signature data to maintain accuracy.

In conclusion, failure signatures are indispensable for effective failure prediction and predictive maintenance. They provide a systematic approach to identifying and responding to potential failures, thereby enhancing equipment reliability and reducing maintenance costs. However, their successful implementation requires high-quality data, robust analytical methods, and consideration of the complexities of modern manufacturing processes.

6.1.5 Data Collection for Signature Development

Data collection is a fundamental step in failure signature development, as it directly impacts the quality and reliability of the failure prediction models. High-quality data

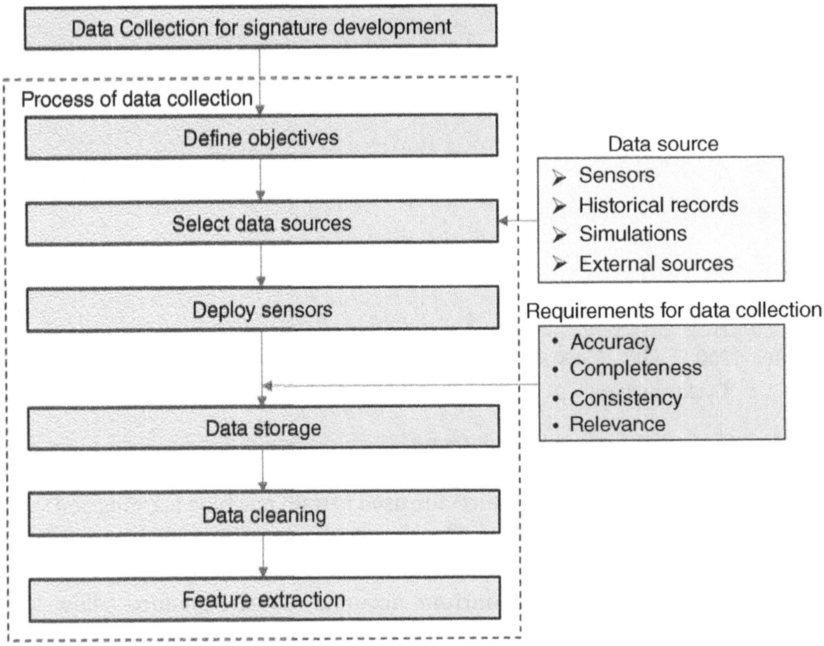

Figure 6.12 Data Collection for Signature Development.

enables the accurate identification of degradation patterns and failure signatures, which are crucial for proactive maintenance and reliability improvement. Effective data collection ensures that the models are well informed, leading to more precise predictions and better decision making.

Figure 6.12 illustrates the collection process of the data, along with common data source and requirements for data.

Data Collection Process

- **Define Objectives:** Clearly outline the goals of data collection, such as identifying specific failure modes or monitoring particular system components.
- **Select Data Sources:** Choose appropriate data sources based on the research objectives. Common sources include sensors, historical maintenance records, and simulation data.
- **Deploy Sensors:** Install and configure sensors to monitor critical parameters of the system. Ensure that sensors are calibrated and maintained regularly to ensure data accuracy.
- **Data Storage:** Store collected data in a structured format, such as databases or data warehouses, to facilitate easy access and analysis.
- **Data Cleaning:** Remove noise, outliers, and missing values from the data to improve its quality and reliability.
- **Feature Extraction:** Identify and extract relevant features from the raw data that are indicative of degradation patterns and failure signatures.

Defining Objectives: The primary goal was to create a repository of vibration signals from normal and faulty bearings, allowing direct comparison of diagnostic techniques across studies.
Selecting Data Sources: A 2 HP Reliance Electric motor, coupled with a torque sensor and loading assembly, served as the test rig. Faults of precise diameters (0.007″–0.040″) were introduced into the inner raceway, rolling elements, and outer raceway via electro-discharge machining (EDM), simulating real-world defect scenarios.
Deploying Sensors: Tri-axial accelerometers were mounted at the 12 o'clock positions on both the drive-end and fan-end housings; in some trials, an additional sensor was affixed to the motor's base plate. Vibration signals were recorded synchronously through a 16-channel DAT recorder, while rotational speed and load data were logged manually.
Data Storage: All measurements were saved in MATLAB (*.mat) format. Each file contains time-series data for fan-end and drive-end accelerometers (and, if used, the base-plate sensor), along with corresponding RPM values. Sampling rates were 12 kHz for general tests and an additional 48 kHz channel for drive-end fault cases.
Data Cleaning: Raw vibration records were imported into MATLAB for de-noising, alignment, and format conversion, ensuring consistency and reliability prior to any analysis.
Feature Extraction: Researchers typically derive a range of indicators from the cleaned signals, including time-domain features (e.g., root mean square, variance, skewness, kurtosis); frequency-domain features (e.g., amplitudes at bearing characteristic frequencies); time–frequency features (e.g., statistics from short-time Fourier transform, Wigner–Ville distribution, wavelet packet decompositions). These features form the basis for subsequent fault classification and prognostics studies.

Data Sources for Failure Signature Development

- **Sensors:** Real-time data from sensors monitoring system performance, such as vibration sensors for mechanical systems or temperature sensors for electronic components.
- **Historical Records:** Past maintenance records and failure logs that provide insights into historical degradation patterns and failure occurrences.
- **Simulations:** Data generated from physical models and simulations, which can be particularly useful when real-world data is limited.
- **External Sources:** Industry benchmarks and research data that can provide additional context and comparison points.

Requirements for Data Collection

- **Accuracy:** Data must be accurate and reliable to ensure the validity of failure prediction models.
- **Completeness:** Data should cover all relevant aspects of system performance to provide a comprehensive view of degradation patterns.
- **Consistency:** Data should be collected consistently over time to ensure that trends and patterns can be accurately identified.

- **Relevance:** Data should be relevant to the specific degradation patterns and failure modes being analyzed.

By following a structured data collection process and ensuring the quality and relevance of the collected data, organizations can effectively develop failure signatures that enhance their predictive maintenance capabilities and overall system reliability.

References

1 Pecht, M. and Dasgupta, A. (1995). Physics-of-failure: an approach to reliable product development[J]. *Journal of the IES* 38 (5): 30–34.
2 Li, J., Deng, Z., Che, Y. et al. (2023). Degradation pattern recognition and features extrapolation for battery capacity trajectory prediction[J]. *IEEE Transactions on Transportation Electrification* 10 (3): 7565–7579.
3 Li, W., Qiu, M., Zhu, Z. et al. (2016). Bearing fault diagnosis based on spectrum images of vibration signals[J]. *Measurement Science and Technology* 27 (3): 035005.
4 Zhao, Z., Yang, J., Kang, Y. et al. (2024). Microcrack monitoring and fracture evolution of coal and rock using integrated acoustic emission and digital image correlation techniques[J]. *Scientific Reports* 14 (1): 8063.
5 Sarawade, A. and Charniya, N.N. (2019). Detection of faulty integrated circuits in PCB with thermal image processing[C]//2019 International Conference on Nascent Technologies in Engineering (ICNTE). *IEEE* 1–6.
6 Khoddam, M., Sadeh, J., and Pourmohamadiyan, P. (2016). Performance evaluation of circuit breaker electrical contact based on dynamic resistance signature and using health index[J]. *IEEE Transactions on Components, Packaging and Manufacturing Technology* 6 (10): 1505–1512.
7 Hu, W., Peng, Y., Wei, Y. et al. (2023). Application of electrochemical impedance spectroscopy to degradation and aging research of lithium-ion batteries[J]. *The Journal of Physical Chemistry C* 127 (9): 4465–4495.
8 Mulpuri, S.K., Sah, B., and Kumar, P. (2023). Unraveling capacity fading in lithium-ion batteries using advanced cyclic tests: a real-world approach[J]. *Iscience* 26 (10).

7

Guidelines for PHM System Implementation

CHAPTER MENU

7.1 Enabling Technologies for PHM

7.1.1 Sensor Technology Selection and Integration

Effective PHM (prognostics and health management) implementation hinges on the strategic selection and integration of sensor systems that can reliably capture critical performance, environmental, and operational parameters. Sensor systems not only serve as front-line data acquisition tools but also influence the overall reliability and cost effectiveness of PHM across the product lifecycle.

Developing and selecting robust sensors are foundational to PHM effectiveness. By designing sensors that are resilient to environmental and operational stressors, rigorously assessing their robustness, and strategically integrating them into PHM frameworks, organizations can significantly enhance system reliability, reduce downtime, and enable predictive maintenance strategies aligned with Industry 4.0 and 5.0 initiatives.

When selecting sensors for PHM, the following considerations are essential:

7.1.2 Developing Robust Sensor Technology and Integration Strategy for PHM

Effective prognostics and health management (PHM) relies on robust sensor technologies capable of consistently acquiring high-quality data across varying operational and environmental conditions. The goal is not only to measure critical system parameters but also to ensure long-term reliability and maintainability throughout the system's lifecycle. This requires a systematic process involving sensor technology development, robustness assessment, and final selection.

7.1.2.1 Sensor Technology Development for PHM

To ensure sensor effectiveness and resilience, sensor technologies must be developed with the following priorities:

- **Durability Under Stressors:** Develop sensor packaging and materials to withstand shock, vibration, temperature extremes, humidity, and electromagnetic interference.
- **Miniaturization and Power Efficiency:** Design compact, lightweight sensors for integration into tight spaces, especially in aerospace, automotive, and industrial electronics applications.
- **Self-Diagnostics and Redundancy:** Embed self-monitoring capabilities to assess sensor health and reduce false alarms. Use redundant sensing where critical data must not be lost.
- **Smart Sensing Capabilities:** Incorporate on-sensor preprocessing, edge analytics, or AI-driven anomaly detection to improve real-time performance and reduce data overload.

7.1.2.2 Conducting Robustness Assessment of Sensors

A structured robustness assessment ensures the selected sensors will perform reliably over time and under varied field conditions. This includes:

- **Environmental Stress Screening (ESS):** Evaluate sensor performance under thermal cycling, humidity, mechanical shock, and vibration.
- **Design of Experiments (DOE):** Use robust engineering or other robust design methods to understand how sensor performance metrics vary with noise factors.
- **Signal Accelerated Degradation Testing:** Analyze signal fidelity over time under accelerated aging to assess long-term accuracy and drift.

Cross-Sensitivity Analysis: Ensure that sensor output is not adversely affected by non-target environmental variables or cross-coupled signals.

7.2 Identifying and Selecting Robust Sensors for PHM

The selection process involves a systematic evaluation using criteria derived from both system-level needs and component-level capabilities:

Selection Criteria	Description
Measurement Relevance	Sensor must detect physical variables tied to failure mechanisms (via FMEA)
Accuracy and Stability	Long-term stability, low noise, and high repeatability are critical
Dynamic Response	Ability to detect fast-changing or transient events (shock, surge, etc.)
Form Factor and Integration Feasibility	Suitability for system constraints (size, weight, mounting)
Survivability	Withstand expected stressors without losing function or calibration
Cost vs. Performance	Evaluate lifecycle cost effectiveness for high-reliability systems

7.3 Integration and Validation for PHM-Ready Systems

Once selected, sensors must be properly integrated into the PHM ecosystem:

- System Architecture Integration:
  - Align sensor placement with failure-prone areas.
  - Design robust communication links (wired/wireless) with error handling.
- Data Validation and Fusion:
  - Validate sensor data streams under real operational loads.
  - Use sensor fusion to combine multiple sensor types for higher fault detectability.
- PHM Algorithm Calibration:
  - Calibrate predictive models using verified sensor data under nominal and degraded conditions.

- Incorporate machine learning techniques to improve anomaly detection and remaining useful life (RUL) predictions.

Hardware-level Integration

a) **Physical Compatibility and Environmental Robustness**
   Sensors must be physically compatible with the host system and capable of withstanding the environmental conditions to which the system is exposed.
   - **Physical Compatibility:** Sensors must be physically compatible with the host system. For example, MEMS/NEMS technologies enable embedding sensors in confined spaces (e.g., PCB-mounted accelerometers in avionics) while maintaining structural integrity [1, 2].
   - **Environmental Robustness:** In harsh environments (e.g., offshore platforms, aerospace engines), sensors must meet standards like IP67 or MIL-STD-810G to ensure resistance to moisture, shock, corrosion, and temperature extremes.

b) **Power Management**
   Long-term operability hinges on smart power strategies:
   - **Energy Harvesting:** Integration of piezoelectric or thermoelectric harvesters enables maintenance-free operation in high-vibration or thermal-gradient environments (e.g., rotating shafts, turbine casings).
   - **Low-Power Design:** Utilizing ultra-low-power integrated circuits and implementing sleep/active duty cycling strategies extends operational life while ensuring data capture fidelity in intermittent or event-driven scenarios.

c) **Signal Conditioning and Noise Mitigation**
   Sensor outputs often require signal enhancement before acquisition:
   - **Onboard Filtering and Amplification:** Onboard pre-processing (e.g., amplification, filtering) minimizes noise interference. For instance, capacitive sensors in high-electromagnetic environments often integrate shielding and differential signal paths.

Communication and Protocol Standardization

a) **Hybrid Data Transmission Architectures**
   Selecting between wired and wireless transmission methods depends on application criticality, data bandwidth requirements, and environmental constraints.
   - **Wired Systems:** Interfaces such as CAN bus, Ethernet, and IEEE 1451 enable high-throughput and low-latency communication—ideal for time-sensitive applications like powertrain monitoring or avionics diagnostics.
   - **Wireless Systems:** For applications requiring mobility or operation in inaccessible areas, low-power protocols such as Bluetooth Low Energy (BLE), Zigbee, or LoRaWAN provide viable options. These are especially suited to distributed sensors on large structures like wind turbines.

b) **Synchronization and Timestamping**
   Precise time alignment across sensor nodes is essential for correct correlation of sensor readings.

- **IEEE 802.1AS (Time-Sensitive Networking):** This ensures deterministic timestamping, enabling accurate phase comparison in vibration monitoring or synchronized load analysis across multiple structural points.
- **Edge-triggered Sampling:** Clock synchronization across channels ensures coherent data collection from multi-modal systems (e.g., concurrent strain and acceleration measurements).

Validation, Calibration, and Fault Tolerance

a) In Situ Calibration

   Long-term reliability demands mechanisms for continuous or scheduled calibration.

   - **Self-calibration Features:** Modern sensors may include internal routines to detect drift or degradation in performance. For instance, pressure sensors can periodically check reference voltages and adjust output to maintain accuracy [3].

b) Redundancy and Fault Tolerance

   Mission-critical applications often require fail-safe mechanisms.

   - **Sensor Redundancy:** Dual or triple redundant sensors combined with voting algorithms reduce false positives and ensure continuity. This is common in nuclear reactors, aircraft control systems, and space missions.

7.4 Advanced Computing Platforms for PHM Analytics

Prognostics and health management (PHM) systems require robust computing architectures capable of ingesting high-volume sensor data, executing complex analytics, and enabling real-time decision-making. Advanced computing platforms provide the performance, flexibility, and scalability required to meet these challenges. This section outlines key computing technologies and their implementation in PHM workflows, drawing from industry practices and academic research. Modern PHM systems are characterized by high-frequency data streams, distributed assets, and the need for fast, accurate prognostics. Advanced computing platforms meet these needs by improving latency, throughput, and model sophistication across the PHM lifecycle. The following are core architectures in PHM deployments:

7.4.1 Edge Computing

Processing data at or near the point of generation—within sensors, controllers, or embedded systems—minimizing latency and network dependency. Edge computing is optimal for safety-critical or remote environments where instant decision-making is vital and connectivity is limited.

7.4.2 Cloud Computing

Elastic, internet-based infrastructure for scalable storage, model development, training, and fleet-wide monitoring. Use cloud platforms for centralized model orchestration, system-wide visibility, and long-term analytics requiring massive computational resources.

7.4.3 Fog Computing

Acts as an intermediate processing layer between the edge and cloud, aggregating and pre-processing data before cloud transmission. Fog computing balances local responsiveness with cloud analytics. It is ideal for moderate-complexity analytics and privacy-sensitive applications.

7.4.4 Distributed Computing Frameworks

Platforms like Apache Spark and Hadoop enable high-throughput, parallel data processing across clusters. Distributed frameworks are well suited for batch processing of large historical datasets, especially in asset-intensive domains.

7.4.5 High-performance Computing (HPC)

HPC are supercomputing clusters designed for simulation-heavy workloads such as finite element analysis and multi-physics modeling. HPC platforms are essential when physics-based models or digital twins form the foundation of the PHM strategy.

7.5 AI-accelerated Hardware

Specialized hardware including graphics processing units (GPUs), tensor processing units (TPUs), and field-programmable gate arrays (FPGAs) is designed to accelerate AI and deep learning tasks. AI accelerators are key for deep learning applications requiring low-latency inference or rapid model iteration.

7.6 Evaluation Metrics for PHM Systems

Evaluating PHM systems requires rigorous metrics that capture both the timeliness and accuracy of degradation detection and RUL estimation. Key metrics include detection time offsets for degradation onset and functional failure, convergence efficiency (χ), prognostic horizon (PH), and α–λ accuracy. This section elaborates these metrics, provides detailed mathematical definitions, and presents calculation examples.

Detection Time Offsets

- Onset of Degradation
  The degradation detection offset, Δt_{deg}, quantifies the delay between the true onset of degradation ($t_{0,true}$) and the algorithm's detected time (t_D):

$$\Delta t_{deg} = t_D - t_{0,true}$$

- Functional Failure Detection
  Similarly, the failure detection offset, Δt_{fail}, measures the difference between detected failure time (t_F) and true end of life ($t_{EOL,true}$):

$$\Delta t_{fail} = t_F - t_{EOL,true}.$$

Convergence Efficiency (χ)

- Convergence efficiency (χ) is defined as the earliest time at which all subsequent RUL predictions remain within an acceptable error bound α:

$$\chi = min \{ t \mid \forall \tau \geq t : |RUL_{pred}(\tau) - RUL_{true}(\tau)| \leq \alpha \}$$

A lower χ indicates quicker stabilization of RUL estimates.

Prognostic Horizon (PH) and α-λ Accuracy

- Prognostic Horizon (PH)
  PH represents the lead time before actual failure when predictions first satisfy the error bound α:
  $PH = t_{EOL,true} - t_P$, *where* t_P *is the first time* $|RUL_{pred}(\tau) - RUL_{true}(\tau)| \leq \alpha$.
- α–λ Accuracy
  α–λ accuracy verifies prediction accuracy at each prediction instance λ within the shrinking error cone:

$$|RUL_{pred}(\lambda) - RUL_{true}(\lambda)| \leq \alpha$$

Example: Consider a component with true EOL at cycle 100. The prognostic model outputs RUL estimates at discrete cycles. Let $\alpha = 5$ *cycles*. The following table lists true and predicted RUL:

Cycle	$RUL_{true}(t)$	$RUL_{pred}(t)$
70	30	40
80	20	27
85	15	19
90	10	13
92	8	10
95	5	4

From the table, the earliest cycle where $|RUL_{true} - RUL_{pred}| \leq 5$ is at $t = 85$ ($|19 - 15| = 4$). Thus: $t_P = 85$, $PH = 100 - 85 = 15$ cycles. To compute convergence efficiency, we check subsequent cycles (≥85): At $t = 90 : |13 - 10| = 3$; $t = 92 : |10 - 8| = 2$; $t = 95 : |4 - 5| = 1 \leq 5$. So $\chi = 85$ cycles.

7.7 Robust PHM System

7.7.1 Modular Architecture for PHM Systems

Data Acquisition (DA)

- **Sensor & network interface:** Collects raw sensor and operational data (e.g., temperatures, pressures, vibrations) from in-flight systems via wired or wireless data-sharing networks.

- **Temporary buffering/storage:** Stores incoming data streams in local buffers or databases for downstream processing and to ensure no loss during network delays or outages.

Data Processing (DP)

- **Signal transformation:** Applies algorithms such as fast Fourier transforms (FFT), digital filtering, and wavelet transforms to convert raw signals into the frequency or time–frequency domain.
- **Averaging and feature extraction:** Performs synchronous/nonsynchronous averaging, statistical computations, and extraction of diagnostic features (e.g., RMS, kurtosis) that characterize machine health.
- **Configured pipelines:** Supports customizable processing chains so different components (engines, gearboxes, bearings) can use tailored algorithms.

Fault Diagnostic Assessment (FDA)

- **Fault detection:** Identifies deviations from normal operating envelopes via thresholding or pattern-recognition techniques.
- **Isolation and identification:** Locates the faulty subsystem and determines the fault type (e.g., crack, imbalance) using model-based or data-driven diagnostic algorithms.
- **Health state output:** Produces discrete or continuous health indices ("health states") that feed into higher-level decision modules.

Prognostic Assessment (PA)

- **Health state estimation:** Uses the extracted features and current health state to model degradation trajectories over time.
- **RUL prediction:** Forecasts the time until a component will no longer meet performance requirements, using model-based, data-driven, or hybrid methods.
- **Failure progression modeling:** Simulates how identified faults will evolve under future operating conditions to support maintenance planning.

Health Management (HM)

- **Integration of FDA and PA outputs:** Consolidates diagnostic and prognostic assessments, along with safety, environmental, and budgetary constraints.
- **Maintenance advisory generation:** Produces recommended actions—such as inspection intervals or part replacements—for an external Maintenance Management System.
- **Decision support:** Enables planners to balance risk, cost, and resource availability when scheduling maintenance activities.

7.7.2 Robustness, Redundancy, and Fault Tolerance in PHM System Design

In the context of PHM, **robustness** is not merely a desirable attribute—it is a **fundamental precondition**. A robust PHM system consistently performs under a wide range of conditions, ensuring long-term reliability, operational confidence, and cost-effective lifecycle management. Achieving robustness requires proactively addressing system variability, degradation, and failure risks from the design phase forward. Among the most critical efforts to ensure this robustness are the **implementation of redundancy** and **fault-tolerant mechanisms** within the PHM architecture itself.

Without robust PHM design, the system tasked with detecting and preventing failures may itself become a point of failure—undermining its entire purpose. Hence, redundancy and fault tolerance are not optional add-ons, but **essential building blocks** for PHM systems that aim to deliver consistent performance and high return on investment.

7.7.2.1 Redundancy in PHM Architecture

Redundancy enhances robustness by ensuring that no single failure leads to a total system breakdown. It builds **resilience** into the architecture and helps validate the integrity of data and decision-making processes.

Key levels of redundancy include:

1) **Sensor-level Redundancy**
   - Multiple sensors measuring the same physical parameter (e.g., temperature, vibration) provide cross-verification.
   - Redundant data streams allow for fault detection and isolation in faulty sensors through correlation or majority voting logic.
2) **Computational Redundancy**
   - Health assessment algorithms run on multiple processors (e.g., dual-core CPUs, distributed nodes) in parallel.
   - If one processor fails or becomes corrupted, the others maintain continuous operation, minimizing data loss or delays.
3) **Data Path Redundancy**
   - Multiple, independent channels for communication and data storage (e.g., cloud backups, mirrored databases).
   - Protects against data transmission failures, bandwidth congestion, or localized memory issues.

7.7.2.2 Fault Tolerance Mechanisms

Fault tolerance is the **capability of a system to continue operating correctly even in the presence of faults**. In PHM systems, this ensures continuous delivery of accurate health insights even as components age or encounter transient errors.

Core fault-tolerant strategies include:

1) **Self-diagnostics and Health Monitoring**
   - The PHM system must be self-aware—continuously checking the health of its sensors, software, and communication channels.
   - Anomalies (e.g., data dropouts, algorithm instability) trigger alerts and automated responses.
2) **Failover Logic**
   - Built-in routines automatically reroute tasks to backup components or systems when primary ones fail.
   - Eliminates downtime and avoids manual intervention, which may be delayed or error-prone.

3) **Error Correction and Data Validation**
   - Implements real-time integrity checks, such as CRCs (cyclic redundancy checks), hash functions, and comparison logic across redundant units.
   - Ensures data fidelity, preventing corrupted information from leading to inaccurate diagnostics or prognostics.

7.7.2.3 Building for Long-Term Reliability and Cost Effectiveness

Designing redundancy and fault tolerance into a PHM system is **not a luxury—it is an investment in operational certainty**. These elements directly enhance system robustness by:

- **Mitigating failure risks** at both the sensor and system level.
- **Ensuring uninterrupted performance** in safety-critical or mission-critical environments.
- **Reducing total lifecycle costs** by preventing undetected failures, false alarms, and unnecessary maintenance.

Ultimately, robust PHM systems deliver consistent value by sustaining performance under unpredictable, noisy, and evolving real-world conditions. They turn insight into action—not just once, but reliably over time.

7.7.3 User-centric Design for Ease of Integration

PHM systems aim to accurately detect and diagnose system faults and predict failure times to enable efficient maintenance scheduling. However, the current design of PHM systems often overlooks user factors, leading to limited utilization of these systems by maintainers. UCD, which focuses on meeting user needs and improving system usability, can significantly enhance the acceptance and effectiveness of PHM systems.

Necessity of User-centric Design in PHM

User Needs and Expectations: End-users of PHM systems, such as maintenance personnel and operational managers, have specific needs and expectations regarding system usability, reliability, and information presentation. As stated in [5], "The reliability of the information presented is a critical factor in the user acceptance and trust in a system." By adopting UCD, engineers can better understand and address these user needs, ensuring that the PHM system meets practical requirements.

System Complexity and Uncertainty: PHM systems involve complex data processing and modeling, which can result in uncertainties in fault detection, diagnosis, and prediction. UCD can help designers simplify complex information and present it to users in an intuitive and understandable manner. This reduces user cognitive burden and improves their ability to make informed decisions based on PHM system outputs.

Human–Machine Interaction: Effective human–machine interaction is essential for the successful implementation of PHM systems. UCD emphasizes the design of user-friendly interfaces and interaction methods, enabling users to easily access and interpret system information. This, in turn, enhances user engagement and trust in the PHM system, as highlighted in [6], who found that "there is a correlation between trust in and usage of automation."

7.7.4 Implementation Measures of User-centric Design in PHM

User Requirements Analysis: Conduct thorough research to identify user requirements, including functional, usability, and reliability needs. Methods such as interviews, surveys, and field observations can be employed to gather user feedback and gain insights into their workflows and challenges.

Participatory Design: Involve end-users in the design process to ensure that the PHM system aligns with their actual needs and work practices. This can be achieved through workshops, prototype testing, and iterative design cycles, as suggested by ISO 9241-210 (2010) [7], which states, "Users are involved throughout design and development."

Intuitive User Interface Design: Design a user interface that presents information clearly and intuitively. Use visual elements such as charts, graphs, and color coding to help users quickly understand system status and fault information. Ensure that the interface is easy to navigate and that critical information is prominently displayed.

Customizable Information Presentation: Allow users to customize the information presented based on their specific needs and preferences. This can include adjustable thresholds for fault alerts, customizable dashboards, and selective display of diagnostic details. By providing users with control over information presentation, UCD can enhance user satisfaction and system usability.

System Transparency and Feedback: Ensure that the PHM system provides transparent information about its decision-making processes and uncertainties. Display confidence intervals for predictions and explain the reasoning behind system recommendations. Additionally, provide users with feedback mechanisms to report issues or suggest improvements, as recommended in [8], who emphasized the importance of "feedback to the human operator in terms of making the base information... visible and providing understandable outputs to the operator."

7.8 Robust Prototype and Test-Bench Development for PHM System Validation

Effective validation of a PHM system demands more than theoretical modeling—it requires **robust, hands-on, iterative test environments** that simulate real-world operating conditions and variability. Developing prototypes and test-benches grounded in robustness principles is essential to ensure consistent performance across environmental stresses, system variations, and long-term use.

A **robust PHM validation approach** guarantees that system components—sensors, data pipelines, algorithms, and decision logic—function reliably under uncertainty, enabling confidence in field deployment and lifecycle management.

7.8.1 System-level Requirements with Robustness in Mind

The design process begins by defining **system-level robustness objectives**, considering not only nominal operating conditions but also variability due to aging, wear, environmental extremes, and user behaviors.

Key elements include:

- **Failure modes of interest** (e.g., fatigue, corrosion, thermal cycling, sensor drift).
- **Operational profiles and boundary conditions**, including environmental ranges (temperature, humidity, vibration) and usage variability.
- **Critical system assets:** mechanical (e.g., pumps, gearboxes), electrical (e.g., motors, inverters), or hybrid systems.
- **Robustness goals**: tolerance to input noise, consistent diagnostic performance over time, adaptability to system degradation patterns.

These requirements shape the hardware and software architecture, focusing on **designing for variation** and ensuring performance stability.

7.9 Modular, Robust PHM Prototype Architecture

Modular Prototype Architecture

After the system requirements are defined, we design the PHM system modules based on these system requirements. A modular design approach enables flexible integration of data acquisition and preprocessing unit, and PHM algorithms. According to Figure 7.1, the architecture typically consists of:

- **Data Acquisition and Preprocessing Unit:** Converts analog signals to digital format and applies filtering, normalization, or feature extraction. Design aspects include sensor selection based on expected failure modes (e.g., accelerometers for vibration, thermocouples for temperature). Signal conditioning circuits to amplify or filter noise. Sampling rate and resolution sufficient for high-frequency phenomena such as bearing faults.
- **Diagnostic and Prognostic Core:** Hosts algorithms—model-based, data-driven, or hybrid—that perform fault detection, isolation, and RUL estimation. Design considerations include algorithm latency, memory footprint, and model training needs.
- **Health Management Module:** The module consolidates diagnostic and prognostic assessments, taking into account safety, environmental, and budgetary constraints. It generates specific recommendations for an external Maintenance Management System, such as optimal inspection intervals and part replacements. This tool empowers planners to balance risk, cost, and resource availability when scheduling maintenance activities, ensuring efficient and effective maintenance planning.
- **Communication Interface:** Ensures compatibility with external monitoring systems or edge/cloud-based platforms for real-time PHM feedback. Using standardized interfaces (e.g., CAN, Ethernet, OPC-UA) enhances scalability and interoperability.

7.10 Test-Bench Design for Robustness Validation

The PHM prototype must be validated against diverse test conditions that emulate real-world complexity, including:

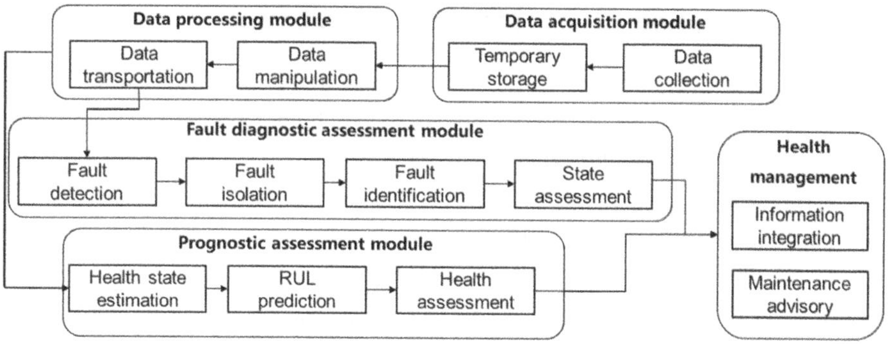

Figure 7.1 PHM Architecture. *Source:* Li et al. [4] / with permission of Elsevier.

- System noise, usage variability, and environmental stressors.
- Sensor degradation modes (e.g., drift, intermittency, noise).
- Failure interactions between subsystems or cascading faults.

A robustness-focused validation approach includes:

- Repeatability and consistency testing: Ensures performance under repeated trials with randomized inputs.
- Stress testing and fault injection: Verifies resilience to data anomalies, communication delays, or sensor failures.
- Comparative benchmarking: Uses controlled degradation profiles to validate diagnostic and prognostic accuracy.
- Robustness assessment for critical components and/or critical subsystems: Evaluates the ability of key elements—such as bearings, actuators, power electronics, and communication modules—to maintain functional integrity under worst-case environmental and operational conditions. This involves testing the limits of component performance while monitoring for deviations in health indicators and diagnostic accuracy.

In addition to physical test-bench experiments, simulation tools can significantly enhance early-stage validation and reduce costs.

Simulation platforms such as MATLAB/Simulink, ANSYS, or Python-based frameworks allow teams to model component behavior, sensor signal characteristics, and system degradation trajectories before committing to full-scale prototyping. These tools enable:

- Rapid algorithm development and debugging.
- Scenario testing under various operational and failure conditions.
- Sensitivity and uncertainty analysis to inform robust design choices.

Model-based validation complements physical testing by exposing edge cases and accelerating system iteration cycles. This dual approach—simulation plus hands-on testing—strengthens confidence in PHM system robustness and ensures reliable operation in dynamic, real-world environments.

7.11 Embedding Robustness into PHM Prototyping

Robust prototype and test-bench development transforms PHM from a conceptual framework into a field-ready solution. By focusing on system variability, environmental stressors, sensor integration, and algorithmic reliability, engineers can:

- Ensure PHM system dependability across lifecycle stages.
- Reduce false alarms and missed detections.
- Optimize cost effectiveness and operational confidence.

Robustness is not a result—it is a design commitment. Through systematic, real-world validation, PHM systems can deliver predictive intelligence with the consistency required for modern industrial and mission-critical environments.

7.12 Verification Against Real-World Failure Data

Even the most advanced PHM systems are only as good as their performance in real-world conditions. While lab-based testing with controlled fault injection and synthetic scenarios is essential for early validation, it does not fully capture the complexity and unpredictability of operational environments. Verification against real-world failure data is therefore critical to ensure that the PHM system is not just theoretically sound, but also practically reliable and deployable.

7.12.1 Why Real-World Data Validation Matters

Real-world operational data includes a range of uncontrolled variables that can significantly affect PHM performance:

- **Sensor drift** and environmental interference degrade signal quality.
- **Multiple overlapping failure modes** can mask or mimic one another.
- **Varying operational conditions** (load, duty cycles, ambient conditions) make fault signatures less distinct.
- **Imperfect ground truth**, where failure events may not be timestamped precisely or diagnosed accurately.

These factors introduce **uncertainty and noise** that lab tests often fail to replicate. Without validation on real-world data, PHM systems risk:

- **Overfitting to ideal conditions** and underperforming in service.
- **High false alarm rates**, leading to unnecessary inspections or part replacements.
- **Missed or late detections**, undermining trust in the system.

Verifying on real-world datasets ensures that the system can withstand data imperfections, maintain accuracy under variable conditions, and integrate into existing maintenance workflows.

7.12.2 Types and Sources of Real-World Failure Data

The credibility of PHM validation depends on the quality and relevance of the real-world data used. Typical sources include:

Maintenance Records

- These provide a log of observed failures, component replacements, and maintenance actions.
- Often include timestamps, fault codes, root causes, and corrective actions.
- Limitation: Records may be incomplete, inconsistent, or delayed in reporting.

Condition-based Monitoring (CBM) Logs

- Continuous sensor readings such as vibration, temperature, pressure, or current over months or years.
- Enable analysis of long-term degradation trends.
- Often synchronized with operational events (start-up, shut-down, load changes).

Incident Reports and Root-Cause Analyses

- Generated after significant failures or unscheduled downtimes.
- Include engineering insights, damage assessments, and cause–effect relationships.
- Useful for modeling rare but critical failure scenarios.

7.12.3 Public Benchmark Datasets

Widely used in research and early-stage model training. When using datasets, it is essential to match their failure mechanisms, operational context, and sensor configuration to the target system to ensure relevant conclusions.

7.12.4 Structured Methods for Real-World Verification

Real world verification should follow a structured methodology to ensure consistent, objective performance assessment. Key approaches include:

a) **Offline Benchmarking**
   - The PHM system is applied retrospectively to historical data with known outcomes.
   - Used to measure detection accuracy, prediction error, and false positives/negatives.

b) **Cross-domain and Cross-system Generalization**
   - A PHM model trained on data from System A is evaluated on System B.
   - Reveals whether the model can **generalize** across asset variants or operational regimes.
   - Useful for fleet-wide systems with variations in design or usage patterns.
   - Techniques like transfer learning and domain adaptation may be employed to boost performance.

c) **Field Trials in Shadow Mode**
   - The PHM system is deployed on active equipment but operates in "shadow mode" providing diagnostics and prognostics without influencing actual maintenance decisions.

- Enables direct comparison with existing maintenance practices: Did the system detect issues that went unnoticed? Would earlier intervention have reduced unplanned downtime?
- Feedback from maintenance personnel is critical to interpret the results and build confidence.

d) **Backtesting Maintenance Impact**
- Historical PHM outputs (real or simulated) are compared with actual maintenance outcomes.
- What-if analysis evaluates: How different alerts would have changed inspection schedules. Potential cost savings from earlier detection or optimized part replacement. Reduction in downtime or asset damage from proactive maintenance.
- Useful in building business cases for PHM adoption and ROI estimation.

Verification against real-world data should not be a one-time step. Instead, it should be integrated into the broader PHM system lifecycle:

1) Model development and tuning using clean datasets.
2) Internal validation on controlled or simulated faults.
3) External validation on field data from actual assets.
4) Continuous learning: Update models as new failure cases and operational data become available.

This closed-loop development ensures that PHM systems evolve with the assets they monitor, improving accuracy and trust over time.

7.12.5 Continuous System Evaluation Post-deployment

Deploying a PHM system into an operational environment is not the end of the validation process—it is the beginning of continuous performance assurance. Over time, changes in equipment condition, operational context, and environmental stressors can degrade the accuracy and relevance of PHM outputs. Continuous system evaluation post-deployment ensures the PHM system maintains reliability, adapts to shifting conditions, and delivers actionable insights consistently.

7.12.6 Rationale for Continuous Evaluation

After deployment, several factors may affect system behavior:

- **Equipment aging** alters fault signatures and degradation patterns.
- **Sensor degradation or recalibration drift** can introduce bias into measurements.
- **New failure modes** may emerge that were not part of the original training data.
- **Operational changes** (e.g., load profiles, duty cycles, climate exposure) affect baseline behavior.

Without monitoring and recalibration, PHM models can become **obsolete**, leading to **false alarms**, **missed detections**, and **loss of trust** from maintenance personnel.

7.12.7 Key Components of a Post-deployment Evaluation Framework

A robust continuous evaluation framework typically includes the following components:

a) **Performance Monitoring Dashboard**
   - Continuously monitor detection accuracy, false-positive rates, and prognostic errors in live systems. The evaluation metrics for PHM systems outlined in Section 1.6.2 can also be applied.
   - Flags model drift or degradation in prediction quality.

b) **Feedback Loop from Maintenance Actions**
   - Maintenance teams annotate PHM alerts with actual inspection or repair outcomes.
   - Feedback is ingested into the system to confirm whether predictions were accurate, missed, or incorrect.
   - Enables supervised learning on newly labeled field data.

c) **Adaptive Model Updating**
   - Online learning algorithms or retraining pipelines allow the PHM model to incorporate new data periodically.
   - May include: **Batch retraining** (e.g., every 3–6 months) using updated labeled datasets. **Incremental learning** to integrate new patterns without losing past knowledge.
   - Version control and rollback mechanisms ensure only validated models are promoted.

d) **Drift Detection and Alerting**
   - Statistical drift detection techniques (e.g., KL divergence, population stability index) compare live data distribution with baseline training data.
   - Alerts are generated if sensor behavior or failure patterns shift significantly—indicating that retraining is needed.

e) **Digital Twin Integration**
   - If available, a digital twin of the monitored system can be used for ongoing simulation, fault replication, and comparison with actual PHM outputs.
   - Enables proactive detection of model performance issues in simulated environments before real-world consequences occur.

7.13 Organizational Integration and Governance

To ensure sustainable post-deployment monitoring, the PHM system must be embedded into the organization's operational and decision-making structures:

- **Roles and responsibilities** should be clearly assigned for monitoring, updating, and maintaining the PHM system.
- **Data governance policies** must define how sensor data, logs, and feedback annotations are stored, accessed, and protected.
- **Periodic audits or performance reviews** should be conducted to evaluate business impact, including cost savings from reduced unplanned downtime, improvements in part replacement efficiency, and enhanced safety and regulatory compliance.

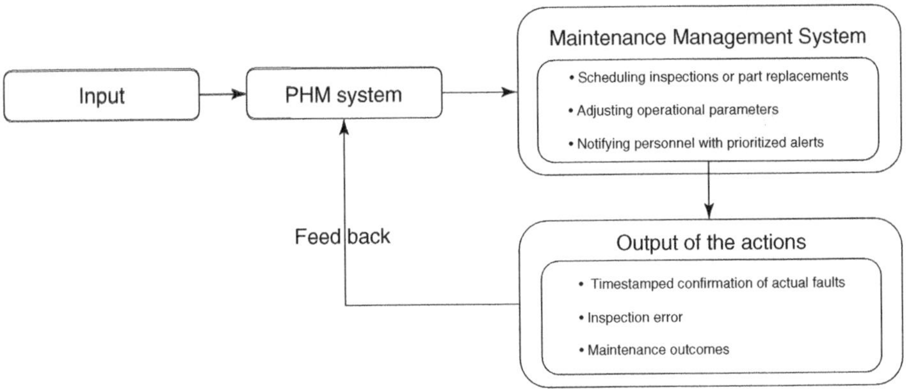

Figure 7.2 PHM-triggered Actions and Data Feedback Loop.

7.13.1 Strategic Implementation of PHM

7.13.1.1 PHM-triggered Actions and Data Feedback Loop

A PHM system's value is realized only when its outputs lead to **decisive, informed actions**. Figure 7.2 shows the triggered Actions and Data Feedback Loop. The detection of a fault or prediction of a degradation trend should trigger a set of coordinated responses, typically routed through a **maintenance management system (MMS)** or a computerized maintenance management system (CMMS). These actions may include:

- Scheduling inspections or part replacements.
- Adjusting operational parameters (e.g., reducing load or cycle time).
- Notifying personnel with prioritized alerts.

Each action taken as a result of PHM outputs generates valuable operational and contextual data. This includes:

- Timestamped confirmation of actual faults.
- Inspection results (found vs. predicted fault).
- Maintenance outcomes (e.g., parts replaced, asset downtime).

This feedback should loop back into the PHM system, enriching the dataset for model validation, retraining, or threshold tuning. Without this closed loop, PHM systems become stale, unable to adapt to real-world variance, and prone to error accumulation.

7.13.1.2 Enhancing PHM Models with Operational Data

Field data offers a dynamic and ever-growing resource for improving PHM models. Once deployed, PHM systems must move from static algorithm execution to **learning from experience**. Key methods for doing this include:

1) **Incremental Learning and Online Updates**
   Algorithms can be updated as new labeled data (e.g., confirmed faults, RUL events) becomes available, avoiding the need for full retraining. This allows models to adapt in near real-time to shifts in equipment behavior or usage patterns.

2) **Domain Adaptation and Transfer Learning**
   Operational data from one machine or environment can help improve model performance across similar platforms, especially in large fleets. Transfer learning enables sharing of insights across related but not identical systems, enhancing fault coverage and generalizability.
3) **Data Augmentation and Simulation**
   When rare or catastrophic failures occur, synthetic examples derived from simulations or generative models (e.g., GANs, physics-informed models) can be combined with operational data to expand the training set and improve detection capability.
4) **Context-aware Modeling**
   Operational data can embed context — like load conditions, mission profiles, or environment — into the model, allowing PHM algorithms to **adapt thresholds dynamically** and reduce false positives during known operational anomalies (e.g., cold starts, high-load periods).

7.13.2 Future-proofing PHM Systems for Technological Advancements

To remain effective over the lifecycle of assets and beyond, PHM systems must be designed for flexibility and scalability. Technology and operational landscapes evolve, and static models or fixed architectures quickly become obsolete. Future-proofing strategies include:

- **Modular and Scalable Architectures**
  PHM systems should adopt a service-oriented or modular design, where components (diagnostic engine, prognostic model, communication layer) can be updated independently without system-wide redeployment.
- **Edge-Cloud Integration**
  Future systems will increasingly rely on edge computing for low-latency fault detection and cloud infrastructure for long-term learning, model updates, and fleet-wide pattern analysis. A hybrid architecture enables real-time local actions while leveraging centralized intelligence.
- **Cybersecurity and Data Integrity**
  As PHM becomes more interconnected, ensuring secure communication, data integrity, and access control becomes essential. Future-ready systems must include encryption, authentication, and anomaly detection to guard against data tampering or cyber attacks.
- **Compliance with Open Standards**
  Building on open protocols and interoperable data formats (e.g., OPC UA, ISO 13374, Digital Twin standards) allows PHM systems to evolve with industry ecosystems and integrate with new platforms, sensors, or AI tools as they emerge.
- **AI Governance and Human-in-the-Loop**
  As AI-driven PHM becomes more autonomous, human oversight remains essential. Future systems must include auditability, explainability, and override capabilities to ensure trust, regulatory compliance, and operational accountability.

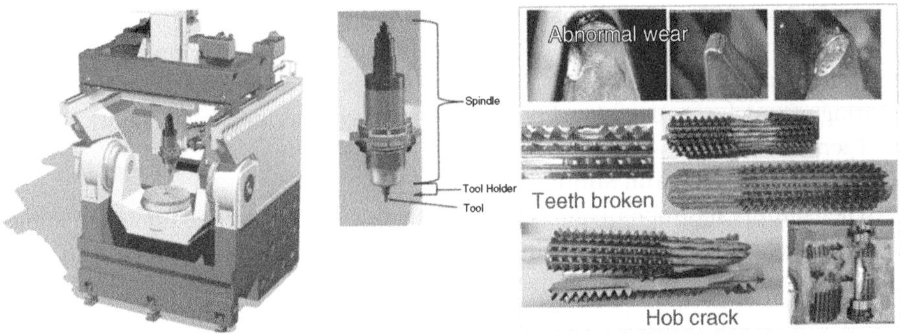

Figure 7.3 Machine Spindles and Cutting Tools Failure. *Source:* Beijing Daolun Technology Co. Ltd.

7.14 Case Study of PHM System Development

The objective is to monitor the operational status of machine spindles and cutting tools to support predictive maintenance (PdM). Figure 7.3 shows the machine spindles, cutting tools and its failure.

Hardware Configuration and Development

Development of Sensor "Black" box and standard machine connection in order to communicate the monitoring data:

- Key 1: Vibration monitoring, data collect and transmission function (hardware development).
- Key 2: 65% cutdown of the unit price (sensors and blackbox).
- Key 3: Compare the actual condition off with normal condition through data analysis (algorithm and software).

Core tasks:

- Sensors selection and delivery:

An independent system that does not connect directly to the machine control system, thereby avoiding additional operational risks. Figure 7.5 shows the framework of the sensors system.

- Figure 7.4 shows the signal collection box ("blackbox") development:

Algorithm Development

After monitoring data collection through our on-site PC station, we analyzed the data and build the algorithm model.

1) **Data Collection and Frequency Analysis**
   - Vibration signal analysis during working and nonworking periods. Figure 7.6 showns the root mean square of the vibration signal.
   - Frequency analysis of six typical timepoints—main changes observed in high-frequency bands (~8 kHz).

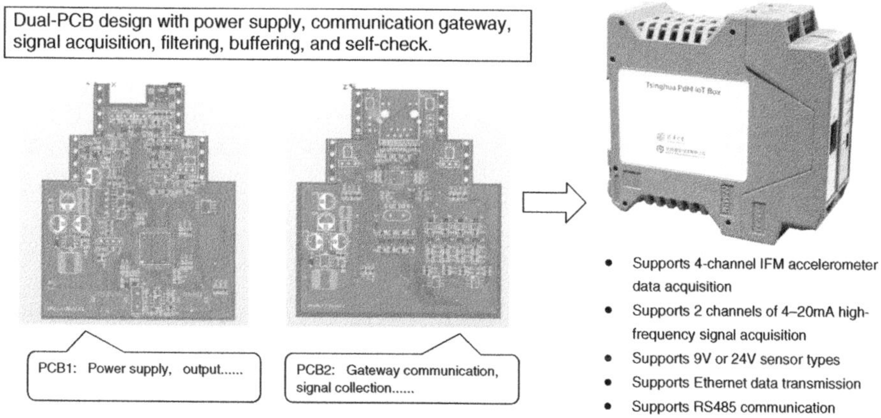

Figure 7.4 Signal Collection Box Design. *Source:* Beijing Daolun Technology Co. Ltd.

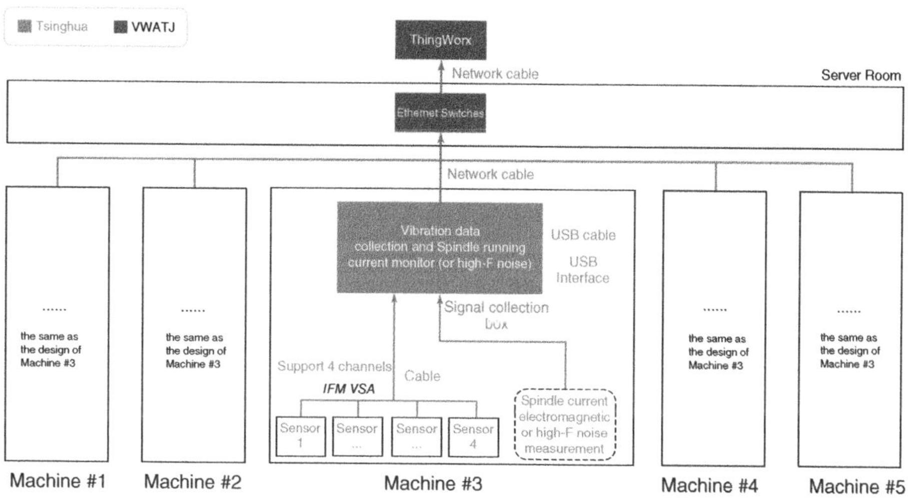

Figure 7.5 Framework of Sensors System.

As shown in Figure 7.7, it obvious that low-frequency spectrum is relatively stable during processing. While the amplitude is large, the frequency spectrum is higher at around 8 kHz. During processing vibration spectrum changes mainly reflected in high frequency area.

- As shown in Figure 7.8, high-pass filtering and envelope spectrum are used to extract characteristic features (e.g., frequency multiplications at 195, 390, 570, 791 Hz).

2) **Time Domain Analysis:**
   - Process duration defined and measured accurately.
   - Figure 7.9 shows the Identification of stable patterns and spike overlap indicating true features.

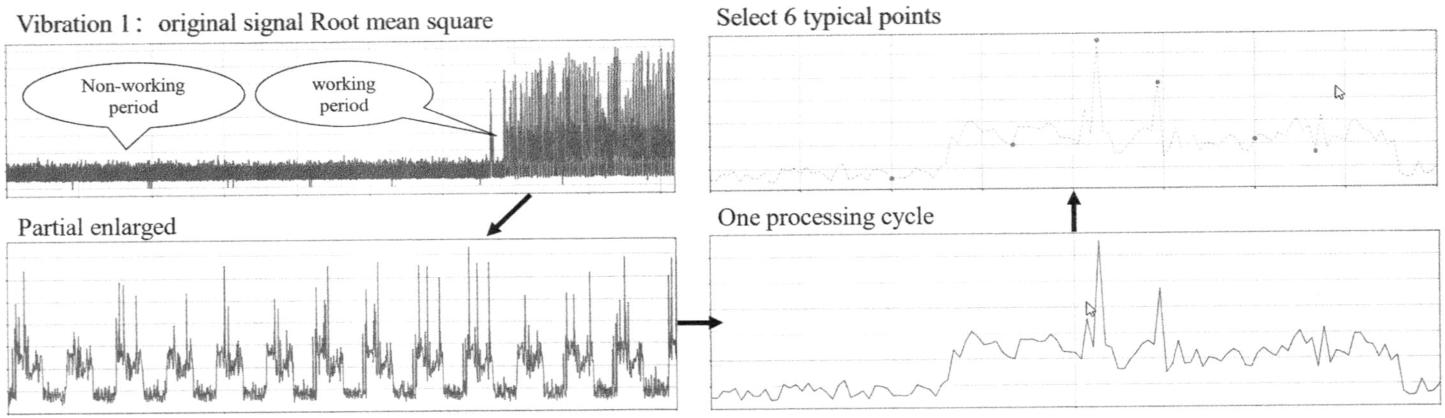

Figure 7.6 Root Mean Square Analysis.

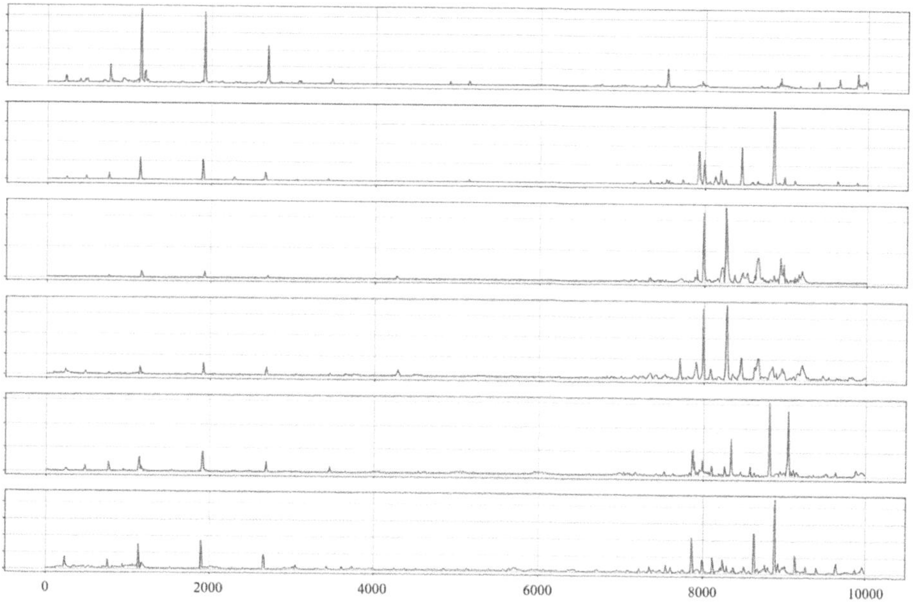

Figure 7.7 Frequency Spectrum of 6 Typical Points.

3) **Health Index Calculation:**
   - Deviation degrees of vibration and current signals quantified.
   - As shown in Figure 7.10, A composite health index is computed to assess machine status.

   After we identified the stable working pattern, we can use it as a standard curve and calculate the deviation of current working curve and standard curve. With the calculation of deviation, we can find the total deviation as the health index. While the health index is significantly low, that means the deviation is huge. The algorithm will alarm to prompt the risk.
4) **Self-learning Mechanism:**

   The alarm limit is based on the historical data training. That means when the machine is working on a different condition with the historical records, the system will remind the operator to find if the machine is working well. If it is well, this record will be stored into the database as a "health" case. Otherwise, it will be recorded as a "failure"case into the database. And next time the similar failure happened, the system will find it and compare with this case and give a note. So "self-learning" process is achieved.

Software Development and Distribution

After the algorithm development, we need to develop and distribute the specific software based on browser/server and the algorithm. This software is capable of:

- Summarized the health index and trend of the monitoring machines, as shown in Figure 7.11.

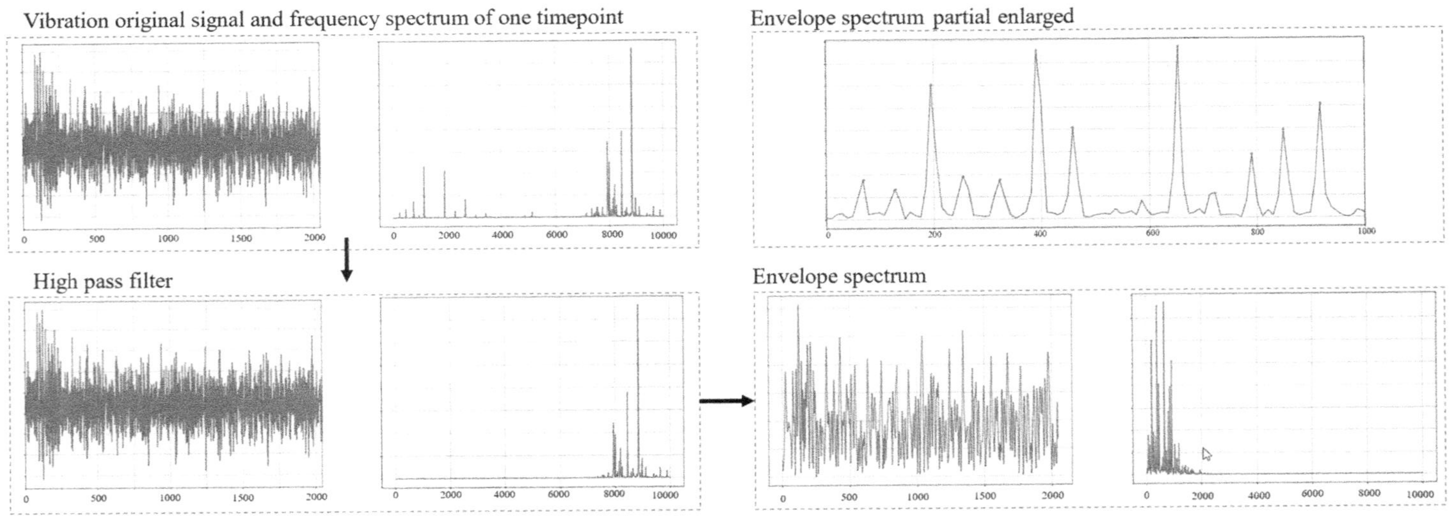

Figure 7.8 High-pass Filtering and Envelope Spectrum.

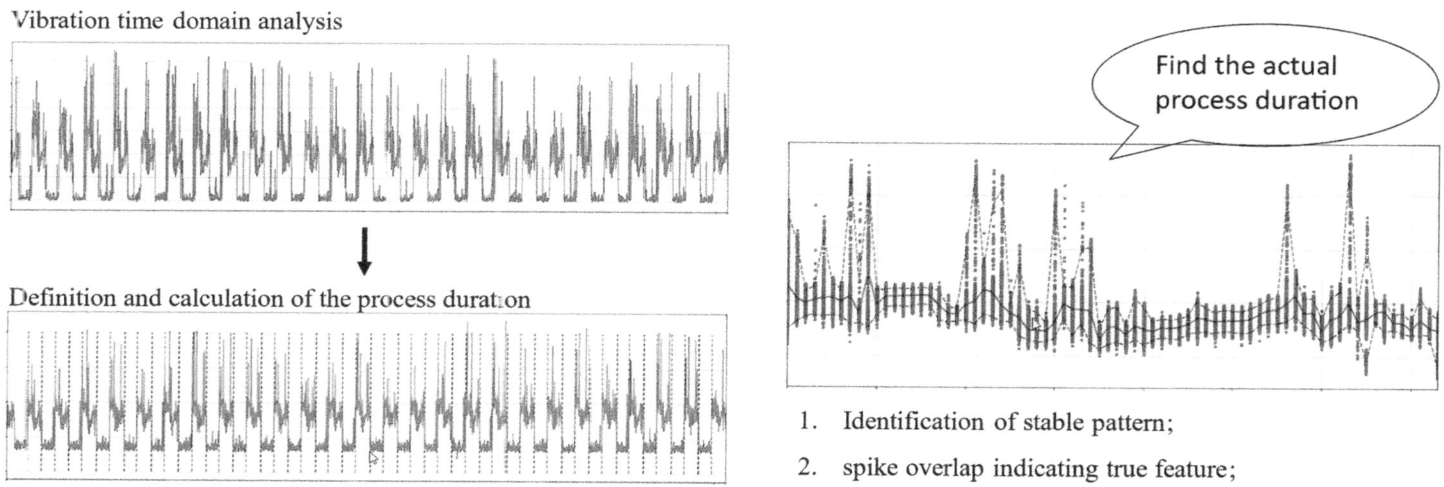

Figure 7.9 Identification of Stable Patterns.

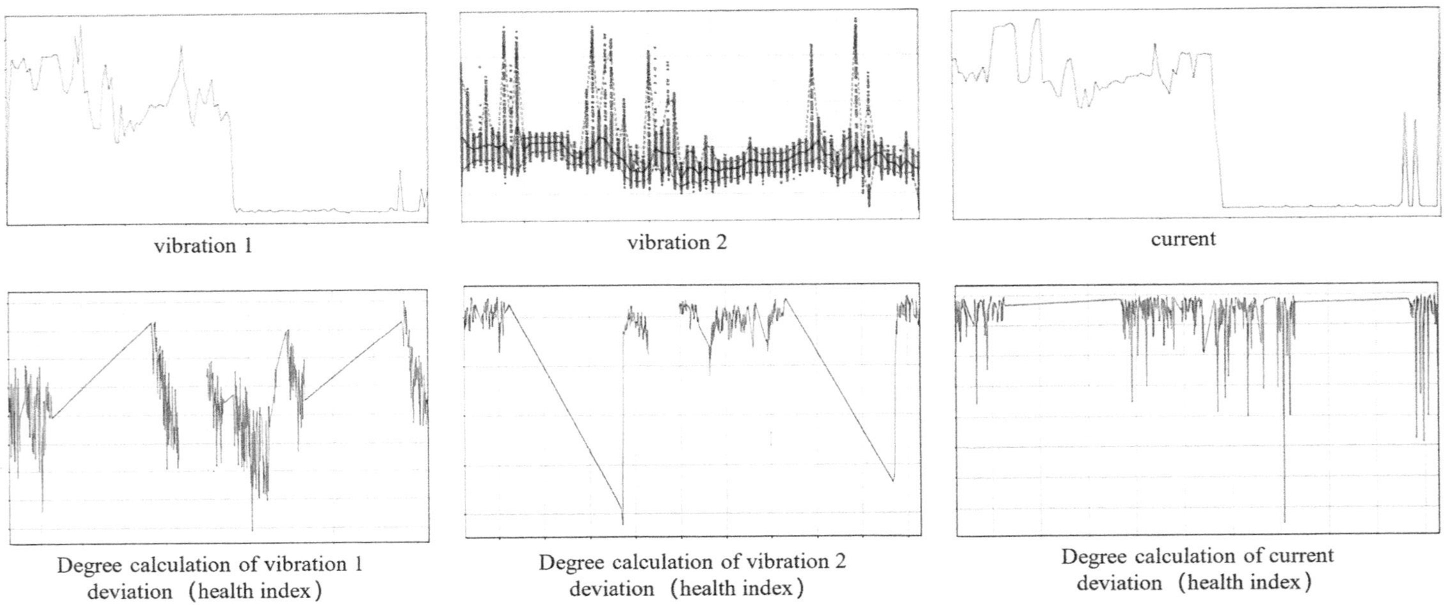

Figure 7.10 Health Index Calculation.

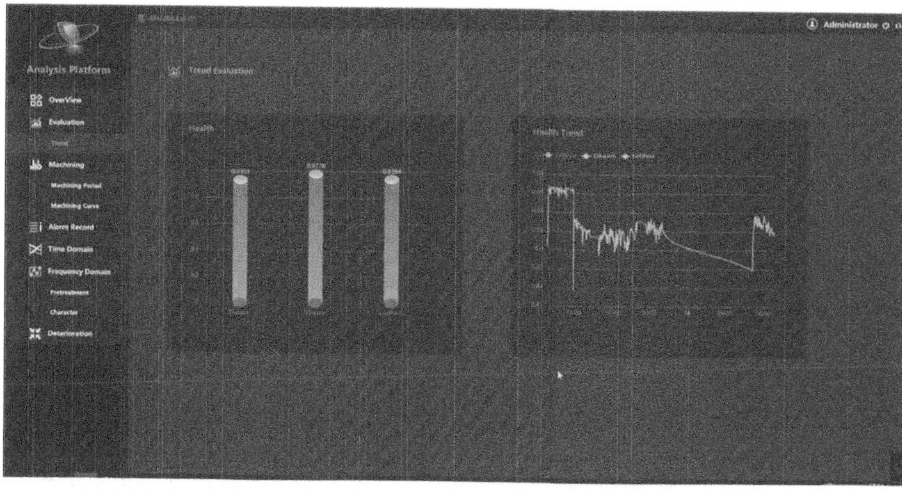

Figure 7.11 Health Index and Trend. *Source:* Beijing Daolun Technology Co. Ltd.

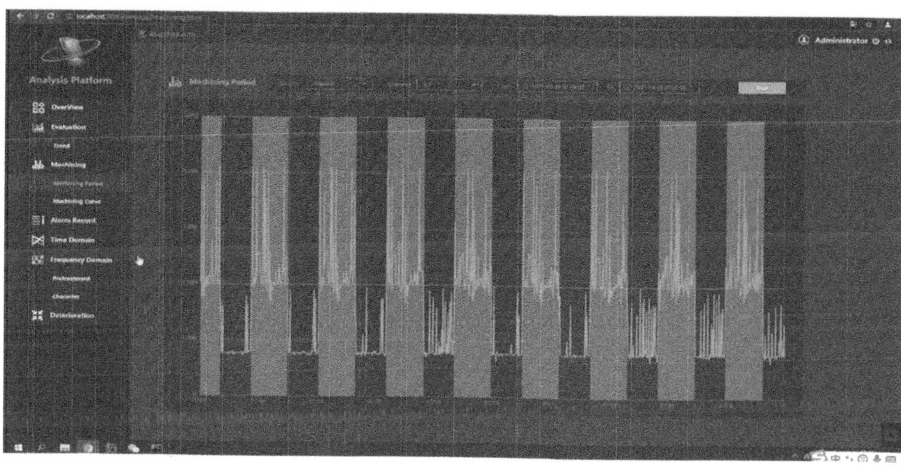

Figure 7.12 Real-time and Historical Signal Curves. *Source:* Beijing Daolun Technology Co. Ltd.

- Visualize real-time and historical signal curves (vibration, acoustic, current), as shown in Figure 7.12.
- Compare actual vs. standard curves, including 5%–95% confidence bands, illustrated as Figure 7.13
- View both time-domain, frequency-domain parameters and the overall condition of the machine. Figure 7.14 shows the deterioration record of the machine.

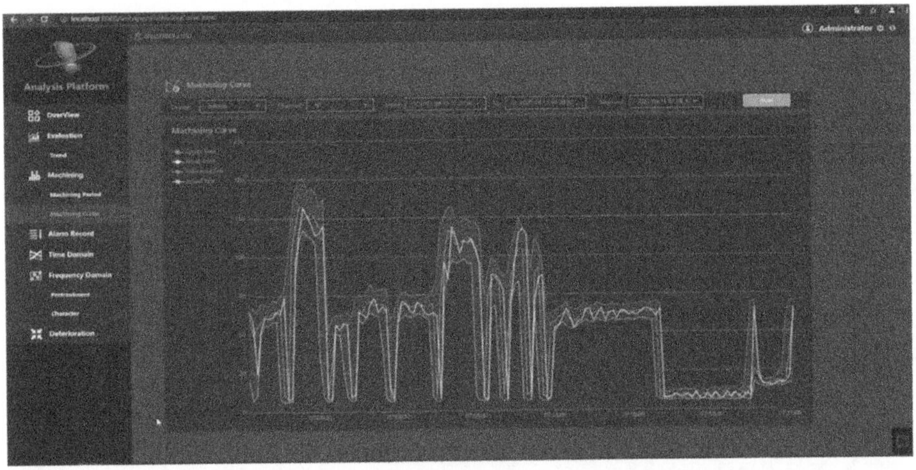

Figure 7.13 Compare Actual vs. Standard Curves. *Source:* Beijing Daolun Technology Co. Ltd.

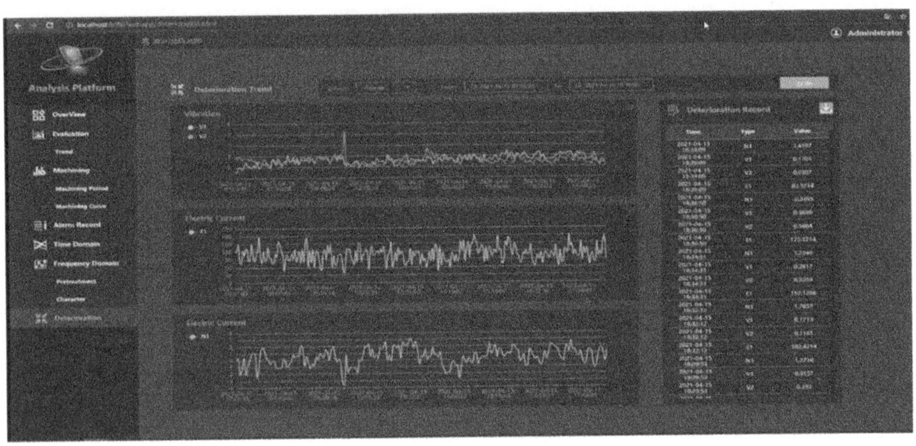

Figure 7.14 Deterioration Record. *Source:* Beijing Daolun Technology Co. Ltd.

References

1 Cheng, S., Azarian, M. H., & Pecht, M. G. (2010). Sensor systems for prognostics and health management. *Sensors*, 10(6), 5774–5797. https://doi.org/https://doi.org/10.3390/s100605774

2 Cheng, S., Azarian, M., & Pecht, M. (2010). Sensor selection for PHM based on failure modes and effects analysis. *IEEE Transactions on Reliability*, 59(1), 131–138. https://doi.org/https://doi.org/10.1109/TR.2009.2036730

3 Fonseca, M. A., English, J. M., Von Arx, M., & Pister, K. S. J. (2006). Integrated pressure sensors with built-in self-test functionality. *Journal of Microelectromechanical Systems*, 15(6), 1608–1617. https://doi.org/https://doi.org/10.1109/JMEMS.2006.880320

4 Li, R., Verhagen, W. J., & Curran, R. (2020). A systematic methodology for prognostic and health management system architecture definition. *Reliability Engineering & System Safety*, 193, 106598. https://doi.org/https://doi.org/10.1016/j.ress.2019.106598
5 McDonnell, D., Balfe, N., Al-Dahidi, S., & O'Donnell, G. E. (2014). Designing for human-centred decision support systems in PHM. In Proceedings of the European Conference of the Prognostics and Health Management Society 2014 (pp. 1–10).
6 Balfe, N. (2010). Appropriate automation of rail signalling systems: a human factors study (Doctoral dissertation). University of Nottingham.
7 International Organization for Standardization. (2010). Ergonomics of human-system interaction – Part 210: human-centred design for interactive systems (ISO 9241-210:2010). https://www.iso.org/standard/52075.html
8 Balfe, N., O'Donnell, G. E., & Lee, J. D. (2012). Automation bias in a rail signalling control task. In Proceedings of the Human Factors and Ergonomics Society Annual Meeting, 56(1), 1–5. https://doi.org/https://doi.org/10.1177/1071181312561001

8

Case Study for Robust Design for Reliability (RDfR)

CHAPTER MENU

8.1 Introduction

As we explore the details of the robust design for reliability (RDfR) framework, process, and scorecard, we will bring these concepts to life through a real-world case study. This case study provides a concrete example of how RDfR methodologies can be effectively applied and implemented in real work environments. By diving into this example, you'll gain a step-by-step understanding of the RDfR framework's application, the challenges faced, and the tangible benefits achieved. This hands-on approach not only clarifies the theoretical aspects but also equips you with a practical roadmap for applying these strategies to your own projects, ensuring enhanced reliability and performance in your work.

Once the business decision is approved—based on the maturity of the business case, initial opportunity analysis report, and the assessment of requirements and risks—a project plan and concurrent team are established, signaling the readiness to initiate the product development effort.

A key advantage of using the RDfR scorecard is its ability to support prognostic health management (PHM) systems [1]. PHM is increasingly vital in industries where continuous monitoring and product health maintenance are essential for ensuring long-term reliability. By identifying key design parameters [2] and critical health indicators early in the design process, the RDfR scorecard aids in developing more effective PHM systems, enabling organizations to predict and prevent potential failures throughout the product's lifecycle. This foresight minimizes downtime and extends the product's operational life, ensuring long-term reliability and efficiency.

Ultimately, RDfR's structured approach, combined with the PHM system integration, provides a robust foundation for both immediate performance improvement and sustained product longevity.

RDfR Case Study: Detonator Power Supply Module (DPSM)

Overview

The detonator power supply module (DPSM) serves as an ideal case study to illustrate the principles and practices of RDfR. Successfully implemented within a renowned and well-established business, this project showcases how RDfR methodologies can be applied to deliver enhanced reliability and robustness in product development. By examining this real-world example, we can derive key lessons about designing products that consistently perform under varying conditions.

RDfR Team Building

As we discussed the RDfR in chapter 4 sections [3]:

- 4.2.1 "Building a Strong RDfR Team."
- 4.2.2 "Concurrent Engineering in RDfR." and
- 4.2.3 "Implementation Readiness for RDfR."

Having a well-prepared RDfR implementation team is crucial for success. The RDfR team is now ready to tackle the challenging tasks ahead, equipped with the necessary skills and knowledge.

The implementation team will follow a meticulously designed roadmap and adhere to a detailed scorecard to ensure a disciplined approach to RDfR. The roadmap provides a clear sequence of steps, milestones, and deliverables, guiding the team through the implementation process. The scorecard, on the other hand, serves as a performance measurement tool, helping us track progress, identify areas for improvement, and maintain focus on our objectives.

By adhering to these structured frameworks, we aim to systematically integrate RDfR principles into our processes, ensuring that we achieve the highest standards of reliability and robustness in our designs. This disciplined approach not only enhances our ability to manage the complexities of RDfR but also fosters a culture of continuous improvement and operational excellence within the team.

Why DPSM Was Chosen

The DPSM was selected for this case study because it operates in a highly demanding environment where reliability and safety are critical. It is subjected to harsh external conditions, including fluctuating power demands and environmental stressors, making it an excellent candidate for showcasing RDfR methodologies. Additionally, the module's widespread use in critical applications made it vital for the company to ensure both operational robustness and customer satisfaction.

RDfR Scorecard as a Framework

In this case, the RDfR scorecard was employed as a structured framework to guide the engineering team throughout the entire product development process, from concept to final product release. The scorecard provided a clear, proactive approach for addressing potential reliability challenges early, thereby mitigating risks before they manifested as product failures [4].

Key Elements of the RDfR Case Study

1) Project Identification:
   - The DPSM's complexity, reliability, and critical function in detonator systems made it a prime candidate for applying RDfR.
   - A cross-functional team was assembled to ensure diverse perspectives in addressing the DPSM's reliability challenges.
2) Product Development Process:
   - RDfR was integrated into the DPSM's development stages through the identify, design, optimize, and verify (IDOV) phases, aligning with the company's existing product development process [5].
   - The team used predictive and analytical tools to anticipate failure modes and design the product to be resilient against them.
3) Scorecard Implementation:
   - **Identify Phase:** During the early stage, customer needs, operational environments, and potential noise factors were defined. The scorecard helped prioritize reliability goals.
   - **Design Phase:** The RDfR scorecard guided the development of robust design solutions, including detailed failure mode and effects analyses (FMEAs), to systematically identify and prevent potential failure modes.
   - **Optimize Phase:** Critical parameters were optimized in the presence of noise environment to ensure the DPSM performed reliably under a range of environmental conditions.
   - **Verify Phase:** Comprehensive testing, guided by the RDfR scorecard, validated the module's performance in real-world scenarios.

(Continued)

(Continued)

4) Handling Uncertainties and Challenges:
   - The DPSM project had to address unpredictable user environments, power fluctuations, and external noise factors. RDfR provided the tools to mitigate these uncertainties through robust design choices, including noise factor management, analysis, and failure prevention techniques like poka-yoke.
5) Results:
   - By incorporating RDfR methodologies, the DPSM achieved a significant reduction in field failures and an increase in operational robustness.
   - Customer feedback indicated higher levels of satisfaction due to the product's consistent and reliable performance, even in challenging environments.

The DPSM case study highlights the power of RDfR in delivering reliable, robust products. By using the RDfR scorecard as a structured tool throughout the product development lifecycle, the team was able to proactively address challenges, reduce failure rates, and ensure high performance in real-world conditions. This approach not only saved costs associated with field failures but also reinforced customer trust in the product's dependability.

8.2 RDfR Phases in DPSM Case Study

8.2.1 Identify Phase

The Identify phase encompasses both the identification and definition processes, including the Define stage. This phase holds significant importance in establishing the framework for reliability engineering throughout the product development lifecycle. It involves a thorough examination and clarification of customer requirements, acting as the cornerstone for subsequent reliability efforts.

The major deliverables are summarized in Table 8.1

Based on the scorecard major deliverables, the DPSM Case Study Objectives are:

- Improve design and field reliability.
- Optimize the DPSM design in the presence of uncontrollable factors to improve design robustness at the lowest possible costs.
- Identify key design parameters.
- Improve customer satisfactions.
- Investigate NTF (no trouble found) and develop SOP for both product engineering and manufacturing engineering.
- No need for the PHM for this DPSM.

1) **Provides Sufficient Energy for Pellet Detonation:** The DPSM is engineered to deliver precise and adequate energy to ensure the reliable detonation of pellets. This function is crucial for applications where controlled and predictable detonation is necessary for operational success and safety.

Table 8.1 The Major Deliverables Are Identify Phase.

IDOV Phase	Gate	Objective	Major Deliverables
Identify (I)	**Gate 1: Avoid failure modes due to bad planning and definition**	Ensure that customer needs and operational noise factors are well understood to prevent planning-related failures	• **Customer Requirements Plan:** Includes quality history, new, unique, and difficult (NUD) challenges, Kano model, functional targets, reliability requirements, and operational noise • **Reliability Requirements Definition:** Establish reliability goals based on user needs and environmental challenges • **Prognostic Health Management (PHM):** Identify potential PHM needs early in the design process
	Gate 2: Avoid failure modes due to poor formulation and characterization	Ensure the proper formulation of the design to prevent early-stage failures	• **Formulation Plan:** Establish design principles and methods (physics/geometry/materials) to ensure robustness • **Initial Noise Factor Management Strategy:** Define noise factors impacting performance • **Transfer Functions:** Derive key functional relationships between design parameters • **Parsimonious Bill of Materials (BOM):** Optimize material usage while ensuring robustness

2) **Solid Connection from PCB to Electrodes (Solder/Termination Integrity):** Ensuring a robust and durable connection between the printed circuit board (PCB) and the electrodes is paramount. The DPSM is designed with high-quality soldering and termination processes to maintain electrical integrity under various operating conditions. This prevents failures due to poor connections or electrical resistance issues.
3) **Fit in Package:** The DPSM must be compact and designed to fit within a specified package size. This requirement ensures that the module can be integrated seamlessly into larger systems without compromising space constraints or overall system design.

The DPSM is a critical component designed to fulfill three primary functions essential for its application in the field.

The DPSM reliability target requirements are identified and summarized in Table 8.2.

Table 8.2 DPSM System Reliability Target Development.

I. A specified useful life period (5 Years Shelf Life)		Year Shelf Life (date from manufactured (Date/Shift/Code))
II. Definition of failures (soft or hard) to be avoided	Hard	1. Misfire; 2. Unintended fire
	Soft	1. Take too long to fire; 2. Take too much voltage to fire; 3. Diagnostic error
III. The combination of noises to be tested dimensioned over the useful life	a. Manufacturing piece to piece	Manufacturing Process: 1. Capacitor, 2. Spark Gap; 3. EFI: 4. Nona; 5. Supplier subassembly; 6. Product assembly; 7. Gun shop assembly
	b. Aging	1. Storage in bunker; 2. Shipping; 3. Storage in assembled gun
	c. Customer Usage	1. Temperature; 2. Humidity; 3. Shock; 4. Vibration; 5. Chemical attack; 6. Pressure; 7. Radiation
	d. External Environments	Capacitor/Spark Gap, 2. Pellet to Booster and ……
	e. System Interaction/ Interfaces	1. Capacitor/Spark Gap, 2. Pellet to Booster and ……
IV. Required performance (Safety and dependability, high confidence, cost of ownership), ideally from benchmarked competition expressed with a robustness demonstration metric and measuring a distance		1. Unintended fire is not acceptable 2. The likelihood (probability) that there is a 99.34% chance or greater that the system will detonate within 3 seconds with required input range volts without a misfire is at least 90% in the presence of combined noise factors 3. This required performance is based on the benchmark reliability performance of 97.26% up to 2006

8.2.2 Design Phase

The Design phase in RDfR is a critical step aimed at developing products that consistently perform well under a variety of conditions and withstand the test of time. Robust design focuses on creating solutions that are not only functional but also resilient to variations in the manufacturing process and environmental factors. The Design phase involves several key activities, which are listed in Table 8.3.

Developing a robust concept is a crucial first step toward achieving overall system robustness. This foundational phase ensures that the core ideas and principles underlying the system are well defined and resilient to various challenges. However, to further enhance the robustness of the entire system, it is essential to move beyond conceptual robustness and engage in comprehensive robust optimization.

The DPSM concept generation and selection is summarized in Table 8.4.

Table 8.3 Identify (I) Phase: Scorecard.

Design (D)	**Gate 3: Avoid failure modes due to poor technical execution**	Prevent technical failures caused by poor execution and alignment of the design with its operational environment	• **Robustness Evaluation Plan: Defining Reliability Requirements:** Understanding and translating customer needs and reliability standards into specific, measurable design criteria • **Concept Generation and Evaluation:** Developing and assessing multiple design concepts to identify the most robust and reliable options • **Detailed Design and Analysis:** Creating comprehensive design specifications, including materials, components, and assembly processes, with a focus on reliability and robustness • **Ensure robustness:** Developing robustness and reliability checklist (RRCL) by mapping failure modes to noise factors and planning mitigations • **Failure Mode Mapping:** Identify how noise factors impact design, missing tests and develop a plan to mitigate these failures in RRCL • **Design Failure Mode and Effects Analysis (DFMEA)** and **Process Failure Mode and Effects Analysis (PFMEA):** Establish a detailed analysis of potential design and process failure modes • **Tolerance Design:** Determining acceptable levels of variability in product dimensions and performance attributes to ensure consistent quality • **Critical Parameters Management (CPM):** Identify and select critical-to-quality (CTQ) and critical-to-reliability (CTR) parameters, incorporating them into health indicators for PHM • **Design Review Based on Failure Modes (DRBFM):** Conduct reviews focused on failure-mode avoidance

Table 8.4 Pugh Concept Selection for DPSM.

Detonator Power Supply Module Concept Selection				
		Original Design	**New Concept 1**	**New Concept 2**
1	Meet all FRs (Functional Requirements) and Constraints	Datum	0	0
2	Distinguish Ability		0	0
3	Safety		0	0
4	Robust Technology		0	+
5	Reliability		0	+
6	Design Cost		0	–
7	TCO (Total Cost Ownership)		+	+
8	Number of Components		–	0
9	Concept Robustness		0	+
10	80%–95% reduction in maintenance time		–	+
11	Less push on electrical connections completed during physical assembly		0	+
...	......			
Total	+ (better) Signs		4	6
	– (worse) Signs		4	3
	0 (same)		7	8

8.2.3 Function Structures

After completing the concept design phase, we carefully evaluated and selected the most promising concept for further development. This critical selection process marks the transition from conceptualization to detailed design, where the focus shifts to understanding how the system's functions can be effectively structured and optimized. Function structure development and its cascading are crucial at this stage, as they provide the framework needed to thoroughly analyze how subsystems, components, and individual elements interact and contribute to overall system performance [6]. By decomposing the system into its functional elements, we can identify the relationships and interfaces that are essential for meaningful and targeted analyses.

In Chapter 4, we introduced the boundary diagram as a tool to clearly define the scope of a system, subsystem, or component, establishing the focus of RDfR efforts. Boundary diagrams serve to delineate the system under consideration and identify the critical interfaces that must be managed to ensure robust performance. Building on this foundation, we now extend the use of boundary diagrams to capture not only the structural relationships but also the functional flows between the various subsystems, components, and functions within the system or product.

This extended approach enables us to visualize how energy, material, and information flow through the different functional elements, thereby identifying potential areas of failure, inefficiencies, or risks. By mapping out these functional flows, we can also ensure that the robustness and reliability of the system are not compromised at any stage of the product lifecycle. Furthermore, this detailed understanding of function structures allows for a more refined and targeted application of reliability tools, such as failure modes and effects analysis (FMEA) or fault tree analysis (FTA), ensuring that robustness is embedded from the earliest stages of design through to production and operation.

This functional decomposition and flow analysis are particularly valuable when applying RDfR principles because they offer a systematic approach to dissecting complex systems into manageable, interrelated parts. By breaking down the system into its subsystems, components, and functional elements, we gain a deeper understanding of how each part contributes to the overall system's behavior [7]. This granular insight is essential for identifying potential reliability risks and addressing them early in the design process, well before they manifest as costly failures in real-world applications.

The value of functional decomposition lies in its ability to map out the interactions and interdependencies between various system components. These interactions often introduce potential failure modes that are not immediately apparent when considering the system as a whole. By isolating individual functions and their associated flows—whether they are energy, material, or information flows—we can pinpoint where weaknesses or vulnerabilities might arise. For example, a critical interface between two subsystems might be susceptible to environmental noise or variations in operating conditions, which could compromise the system's overall reliability. Through flow analysis, such risks can be identified, analyzed, and mitigated through robust design adjustments.

Applying RDfR principles in this context ensures that reliability is integrated into the design at multiple levels. At the subsystem level, engineers can focus on robustness in isolation, optimizing the design to handle variability, manufacturing inconsistencies, or external noise factors. Similarly, at the component level, specific design parameters can be optimized to withstand the anticipated stresses and loads during operation. This multi-level approach ensures that reliability is not an afterthought but a fundamental design criterion that permeates every aspect of development [8].

Furthermore, by systematically addressing reliability challenges through functional decomposition and flow analysis, we are able to develop and validate robust design solutions. These solutions are not just theoretical but are grounded in real-world operational contexts, ensuring that the final product can perform reliably under a wide range of conditions. For instance, by simulating different operating environments, such as extreme temperatures, vibrations, or humidity levels, we can test how various subsystems interact and whether any vulnerabilities emerge under stress. This proactive testing allows us to refine the design to be more resilient before it reaches the production stage, thereby reducing the likelihood of field failures.

In addition to improving system performance, this approach contributes to long-term cost savings and efficiency gains. By addressing potential reliability issues early on, organizations can avoid expensive redesigns, recalls, or warranty claims later in the product lifecycle. A product designed with RDfR principles not only functions as intended but also requires less maintenance, has a longer operational life, and enhances customer satisfaction—all of which contribute to a company's bottom line.

Moreover, functional decomposition and flow analysis facilitate a more agile and flexible design process. With a clear understanding of how the different parts of the system interact, it becomes easier to make design adjustments or incorporate new technologies without compromising reliability. This flexibility is particularly important in today's fast-paced manufacturing environment, where Industry 4.0 technologies are pushing the boundaries of product complexity. By embedding RDfR into the design process from the outset, companies can keep pace with technological advancements while ensuring their products remain reliable and robust.

In summary, the functional decomposition and flow analysis process is a cornerstone of applying RDfR principles effectively. It not only provides a clear pathway for addressing reliability challenges at multiple levels but also ensures that the final product meets both its functional requirements and long-term reliability goals. Through this detailed and systematic approach, organizations can design products that are not only innovative and efficient but also robust and reliable in real-world operating environments, delivering sustained value over their entire lifecycle.

In the case study of the DPSM, the focus is specifically on the DPSM itself to explore how detailed functional analysis and robust design techniques can be applied to improve system performance. The functional diagram of the DPSM, presented in Figure 8.1, illustrates the power of leveraging function structures and extending boundary diagrams to capture the intricate relationships between system components. These tools are crucial for ensuring the DPSM not only performs its intended functions but does so with a high degree of reliability and robustness, especially in challenging real-world operating environments.

By providing a clear visual representation of the DPSM's internal functions, the functional diagram enables us to break down the module into its core components and their interactions. This level of detail is critical because it helps to identify potential points of failure or inefficiency within the system that might otherwise go unnoticed. For example, the flow of electrical energy through the power supply, the regulation of voltage levels, and

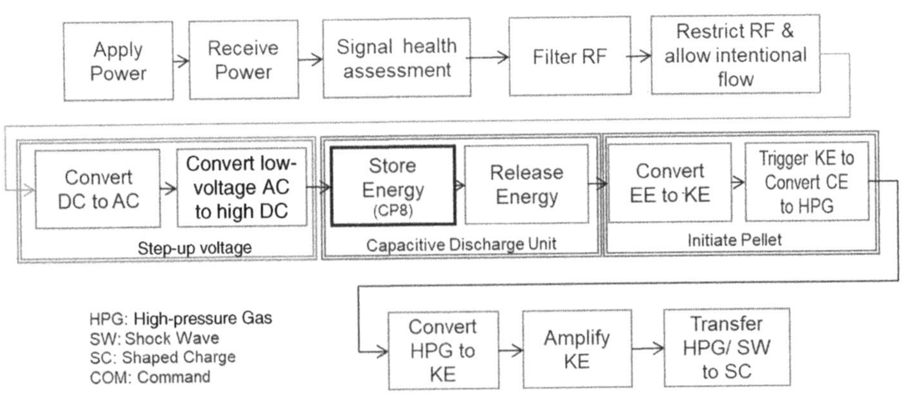

Figure 8.1 DPSM Functional Diagram.

the triggering mechanisms for the detonator all represent key functional areas that must be optimized for reliability. Any weaknesses or inconsistencies in these areas could compromise the overall performance of the DPSM, leading to operational failures in critical applications.

Extending the boundary diagrams to capture the functional flows within the DPSM allows for a more comprehensive system model, which goes beyond just outlining the components. It integrates the relationships between the subsystems and the external environment, considering how the DPSM will interact with surrounding systems and environmental factors. This broader perspective is essential for ensuring the DPSM can perform reliably under varying conditions, such as fluctuations in power input, temperature extremes, and physical stress. By analyzing these interactions through extended boundary diagrams, we can proactively design solutions to mitigate risks and prevent failures.

The integration of functional flows into the system model also facilitates more effective robustness design. Robust design principles emphasize creating systems that are less sensitive to variability in both the manufacturing process and the operating environment. In the case of the DPSM, this means ensuring that the module can deliver consistent power to the detonator, even in the presence of external disturbances like noise, vibration, or electrical surges. By modeling the functional flows and identifying areas where variability could introduce risk, engineers can apply design techniques such as redundancy, insulation, or advanced control systems to improve the module's resilience.

Furthermore, capturing detailed functional flows through these models aids in enhancing the reliability of the DPSM. Reliability is a critical factor in systems like detonator power supplies, where failure could have serious consequences. The functional diagram and extended boundary analysis allow engineers to map out potential failure modes at each stage of the DPSM's operation. Whether it is a failure in the power supply circuit, a malfunction in the voltage regulator, or a misfire in the detonator trigger, each potential issue can be examined in the context of the system's overall functionality. By identifying and addressing these failure modes during the design phase, the team can ensure that the DPSM will function as intended over its expected operational life, reducing the likelihood of unanticipated breakdowns or performance issues.

Additionally, this approach improves the performance of the DPSM by enabling optimization at both the system and component levels. For instance, by understanding the functional relationships between the power supply and the detonator trigger, engineers can fine-tune the energy delivery to ensure the detonator receives the exact amount of power needed for reliable activation, without unnecessary power losses or inefficiencies. This not only improves the performance of the module but also extends its lifespan by preventing overloading or wear on key components.

Ultimately, the use of function structures and extended boundary diagrams in the case of the DPSM exemplifies a powerful method for improving complex system designs. By capturing detailed functional flows and integrating them into comprehensive system models, engineers can significantly enhance the robustness, reliability, and performance of the DPSM. This case study highlights the importance of applying these advanced design methodologies to not only meet functional requirements but also to ensure long-term success in real-world operating environments.

By utilizing function structures and expanding boundary diagrams to include detailed functional flows, we can develop robust and comprehensive system models. This approach offers several advantages:

1) Enhanced Understanding of System Dynamics:
   - The functional diagram provides a clear visual representation of how the DPSM operates, highlighting the various functions and their interactions within the system.
   - By mapping out these interactions, we gain deeper insights into the system's behavior and performance.
2) Integration of Detailed Functional Flows:
   - Extending boundary diagrams to capture detailed functional flows allows us to visualize the movement of materials, energies, and signals (information) through the DPSM.
   - This comprehensive mapping ensures that all critical aspects of the system are considered, leading to more accurate and reliable models.
3) Clear System Boundaries:
   - Defining clear boundaries within the functional diagram ensures that the scope of the DPSM is well understood and isolated from external influences.
   - This clarity aids in focusing RDfR efforts on the most critical aspects of the system.

8.2.4 Reviewing and Matching Functions to Devices

Once the functional diagram is established, the next crucial step involves thoroughly reviewing and matching the identified functions to their corresponding devices or physical forms. This process is vital because it bridges the gap between abstract functional requirements and the tangible components that will fulfill those requirements in the actual system. The goal is to ensure that each function identified in the functional diagram is adequately represented by a physical device or form that can perform or support that function reliably. An example of matching functions to devices is shown in Figure 8.2: Example of DPSM Functions and Flows.

During this review process, the functional diagram acts as a guide to evaluate how well each device meets the necessary performance criteria for its assigned function. By mapping functions to devices, we create a clear picture of how each part of the system contributes to the overall performance and reliability. In addition, this process helps ensure that no critical functions are overlooked or inadequately addressed by the design. It also enables the identification of potential inefficiencies or weaknesses where the chosen device may not optimally meet the functional requirements, prompting revisions or alternative solutions.

8.2.5 Summarizing Main Input and Output Flows

In conjunction with reviewing and matching functions to devices, it is equally important to summarize the main input and output flows for each function, as illustrated in Figure 8.2. These flows represent the transfer of energy, materials, or information between different components of the system. By clearly defining the inputs and outputs, we can ensure that each function is supported by appropriate resources and that the system as a whole operates harmoniously.

Device/Form		Main input flow	Function	Main output
Filter RF	C1	Unintended RF	Filter RF	No RF
	R1	Unintended RF	Restrict RF flow and allow intentional flow and serve as a fuse	Restricted flow
Charge pump	Switch	DC	Convert DC to AC	AC
	Multiplier	AC	Covert low voltage AC to higher DC	DC
Capacitive discharge unit	CP8	DC	Store energy	DC
	SG	DC	Release energy	DC
Initiate pellet	EFI	DC	Convert electrical energy to kinetic energy	Kinetic energy
	Barrel	Initial EFI Flyer position/mass/ energy	Provide room for flyer travel (acceleration)	Final EFI Flyer position/mass/ energy
	Nona Pellet	Kinetic energy	Amplify kinetic energy using chemical energy	High Pressure Gas

Figure 8.2 Example of DPSM Functions and Flows.

The input/output flow summary is crucial for understanding how the devices interact with one another, revealing the dependencies and interfaces between different subsystems and components. For example, a power supply module in an electronic system may have an input flow of electrical energy and an output flow of regulated voltage to other subsystems. Accurately documenting these flows enables designers to better manage the allocation of resources, prevent bottlenecks, and reduce the risk of malfunctions that could arise from mismatches in expected inputs and outputs.

This level of detail not only aids in optimizing the performance of individual components but also enhances the overall efficiency of the system. By identifying areas where input or output flows may need to be adjusted, designers can implement corrective measures early in the design process, avoiding costly rework later in development.

8.2.6 Creating a Robust, Efficient, and Reliable System

The process of reviewing functions and matching them to their corresponding devices, combined with summarizing the main input and output flows, is essential for creating a robust, efficient, and reliable system. Each component must be capable of not only performing its designated function but also interacting seamlessly with other parts of the system. This thorough analysis helps ensure that all elements of the system are aligned with the intended functional goals, reducing the risk of failures due to mismatches between functional requirements and physical implementations.

Additionally, this approach facilitates early detection of potential reliability issues. For instance, if a device is found to have a low tolerance for variability in input flows, designers can introduce robustness measures such as redundancy, improved materials, or alternative configurations to mitigate risks. As a result, the system becomes more resilient to variations in operating conditions and is better able to maintain performance over time, leading to higher overall reliability.

8.2.7 Supporting Effective Communication and Application of RDfR Principles

Another key benefit of this detailed approach is that it supports effective communication among cross-functional teams, including engineers, quality managers, and stakeholders. By clearly documenting the relationships between functions, devices, and input/output flows, team members gain a shared understanding of the system's design and operational characteristics. This transparency helps ensure that all parties are aligned on the goals of the project and can make informed decisions about potential trade-offs, design changes, or improvements.

Moreover, this process reinforces the application of RDfR principles. RDfR emphasizes designing systems that are resistant to variability in both the manufacturing process and the operating environment. By carefully reviewing the functions and matching them to appropriate devices, designers can apply robustness techniques to ensure that each device performs reliably under a wide range of conditions. This might include selecting materials with higher durability, designing for redundancy in critical functions, or implementing control systems that adjust to fluctuations in input flows.

Establishing a functional diagram is only the first step in creating a reliable system. The subsequent process of reviewing and matching functions to devices, along with summarizing the main input and output flows, is essential for ensuring that the system operates as intended. This detailed approach not only enhances system understanding and performance but also supports the effective application of RDfR principles, leading to a system that is robust, efficient, and capable of delivering reliable performance in real-world environments. Through these steps, we lay the groundwork for a design that not only meets its functional requirements but also excels in long-term reliability and operational excellence.

Based on the available functional diagram and the mapping of functions to devices, the team is well prepared to take the next critical step: planning and conducting robust optimization in the **Optimize Phase**. This phase focuses on refining the system's parameters to achieve optimal performance, particularly in the presence of noise factors. **Noise factors** refer to unpredictable variables or disturbances, such as environmental fluctuations, material inconsistencies, or human factors, that can adversely affect the system's performance. Properly accounting for these noise factors is key to ensuring system robustness and reliability in real-world applications.

8.2.8 Understanding Control Factors in Robust Optimization

During robust optimization, one of the most important tasks is identifying and classifying control factors—those design or operational parameters that directly influence system behavior. These control factors can be divided into two critical types based on their interaction with noise factors: **Type 1 Control Factors** and **Type 2 Control Factors**. Each type plays a distinct role in achieving robustness and optimizing them requires different strategies.

8.2.9 Type 1 Control Factor: Interaction with Noise Factor

Type 1 control factors interact with noise factors, meaning that their nominal values have a direct effect on the variability of the system's response to external disturbances. These control factors can either amplify or reduce the system's sensitivity to noise, making them crucial targets for optimization.

- **Reducing Variability:** Since Type 1 control factors directly affect how noise influences the system, adjusting these factors allows engineers to reduce the variability in system performance caused by noise.
- **Example:** Suppose a system includes a control factor A, such as temperature, that interacts with an environmental noise factor like humidity. By adjusting the nominal value of temperature, it is possible to stabilize the system's output, reducing performance variation caused by fluctuating humidity levels.
- **Optimization Strategy:** The goal with Type 1 control factors is to desensitize the system to noise, thereby enhancing its robustness. Engineers achieve this by carefully adjusting the control factor to minimize the impact of noise on the system's response. For example, using statistical tools such as **Taguchi methods** or **response surface methodology (RSM)** can help find the optimal level of a control factor where the effect of noise on performance is minimized.

8.2.10 Type 2 Control Factor: No Interaction with Noise Factor

In contrast, Type 2 control factors do not interact with noise factors, meaning that changing their nominal values will shift the system's performance but would not affect its variability. These control factors are essential for fine-tuning system outputs without introducing additional complexity or increasing sensitivity to noise.

- **Shifting Response without Affecting Variability:** Type 2 control factors allow engineers to adjust the overall response of the system to a desired level without altering its sensitivity to noise.
- **Example:** Consider a control factor X, such as the speed setting of a motor, which does not interact with environmental noise factors like vibration. Adjusting the motor speed will shift the system's performance (e.g., increasing productivity), but it would not affect how the system handles variability from vibration.
- **Optimization Strategy:** For Type 2 control factors, the goal is to adjust the system's performance or output to meet desired specifications without introducing additional noise sensitivity. This ensures that while the performance shifts to the desired target, the system remains stable and consistent. Tools such as **design of experiments (DOE)** can be used to systematically adjust Type 2 control factors to achieve the desired outcome while maintaining robustness.

8.2.11 Tailoring Optimization Strategies for Control Factors

By distinguishing between these two types of control factors, the team can craft more effective optimization strategies to meet both reliability and performance goals.

- **For Type 1 Control Factors:** The primary objective is to minimize the effect of noise on system performance. This requires techniques like **robust parameter design**, where the focus is on finding the ideal settings that mitigate the influence of noise, ensuring the system remains consistent even under varying conditions. This ultimately reduces the risk of product failure or degradation over time due to environmental or operational disturbances.
- **For Type 2 Control Factors:** The focus is on fine-tuning the system's response to meet specific performance targets without increasing variability. Adjustments to Type 2 control factors can help align the system's performance with operational goals (e.g., output, efficiency) without compromising on robustness, making the system more flexible and adaptive.

This tailored approach ensures that both the variability of the system and its output levels are addressed in a balanced manner, resulting in a more resilient, dependable, and optimized system that can withstand real-world conditions.

8.3 Achieving System Robustness through Optimization

The journey from **concept robustness** to **system robustness** is a methodical and iterative process that involves applying robust optimization techniques to enhance both the system's performance and its ability to handle noise. By strategically manipulating Type 1 and Type 2 control factors, engineers can create systems that not only meet functional and operational requirements but also achieve long-term reliability, even in challenging and unpredictable environments.

Through robust optimization, the system can be refined to consistently deliver optimal performance, ensuring that it remains stable and reliable across a wide range of operating conditions. This process not only enhances the system's immediate functionality but also increases its lifespan and reduces maintenance needs by preventing failures caused by noise-induced variability.

In conclusion, by recognizing the distinct roles of Type 1 and Type 2 control factors and optimizing them accordingly, the system's overall robustness and effectiveness can be significantly improved. This structured and nuanced approach to robust optimization ensures that the final product is resilient, reliable, and capable of meeting both current and future demands with consistent, high-level performance.

8.4 Optimize Phase

The Optimize phase in RDfR focuses on refining and enhancing the product design to ensure maximum robustness and reliable performance. This phase is critical for identifying and implementing improvements that mitigate risks, reduce variability, and ensure the product performs consistently under diverse conditions. The key objectives of the Optimize phase include [9]:

1) Fine-Tuning Design Parameters: Adjusting design parameters to achieve optimal performance and reliability, ensuring the product meets all specifications and customer expectations [9].
2) Reducing Variability: Identifying sources of variability in the manufacturing process and product operation and implementing strategies to minimize their impact [9].
3) Enhancing Robustness: Applying advanced design techniques to make the product more resilient to environmental and operational variations [9].
4) Risk Mitigation: Using tools such as Failure Modes and Effects Analysis (FMEA) to proactively address potential failure points and ensure that the design changes enhance reliability.
5) Verification and Validation: Conducting thorough testing and analysis to confirm that the optimized design meets all reliability requirements and performs consistently under real-world conditions.
6) Ensuring Producibility
   - Design for Manufacturability (DFM): In the Optimize phase, ensuring producibility is key. This involves refining the design to simplify production processes and selecting materials and tolerances that align with cost-effective manufacturing methods.
   - Process Capability Analysis: Process capability studies (Cp, Cpk) ensure that manufacturing processes can meet the design's specifications consistently.
   - Incorporating the insights from SNR analysis ensures that the design remains robust even when subject to process variations during production (Table 8.5).

8.4.1 P-Diagram: Linking Robustness and Serving as an Input for DFMEA

A parameter diagram (P-diagram) is a powerful tool that establishes a crucial linkage to robustness while serving as a vital input for Design Failure Mode and Effects Analysis (DFMEA). It provides a comprehensive visual representation of the system's dynamics, identifies critical control and noise factors, and aids in optimizing input–output relationships for enhanced robustness [2]. Simultaneously, it serves as a foundational tool for systematically identifying, assessing, and mitigating potential failure modes within the DFMEA process. This dual functionality makes the P-diagram indispensable for achieving reliable, high-performing systems as shown in Figure 8.3.

Here's an expanded explanation on how a P-diagram achieves these objectives:

Linkage to Robustness

1) **Visual Representation of System Dynamics:**
   - The P-diagram provides a visual representation of the system's functional relationships, showcasing the interaction between inputs, outputs, control factors, and noise factors.
   - By illustrating these interactions, the P-diagram helps in identifying critical parameters that affect system performance and robustness.
2) **Identification of Control and Noise Factors:**
   - The diagram helps distinguish between control factors (which can be manipulated to improve performance) and noise factors (external variables that cannot be controlled but affect the system).

Table 8.5 Optimize (O) Phase Scorecard.

Optimize (O)	**Gate 4: Avoid failure modes due to lack of robustness and mistakes (hard and soft failures)**	**Optimize the design to withstand uncontrollable user environments** and mitigate both hard (physical) and soft (functional) failures. Prevent mistakes in design, manufacturing, and operation	• **Robustness Design Strategy:** Ensure that the product is robust against external noise factors (e.g., temperature, humidity, user behavior) Design the product to be insensitive to such environmental variances while maintaining functionality • **Environmental Noise Management:** Implement design elements that address challenges posed by uncontrollable external noise factors • **Mistake Prevention:** Apply mistake-proofing (poka-yoke) techniques to eliminate operational and manufacturing errors • **Hard Failure Mitigation:** Ensure design robustness to prevent material and mechanical breakdowns • **Soft Failure Mitigation:** Ensure design robustness to prevent performance degradation due to operational variations • **Producibility:** Ensure the design is optimized for manufacturing and assembly, reducing the likelihood of production errors

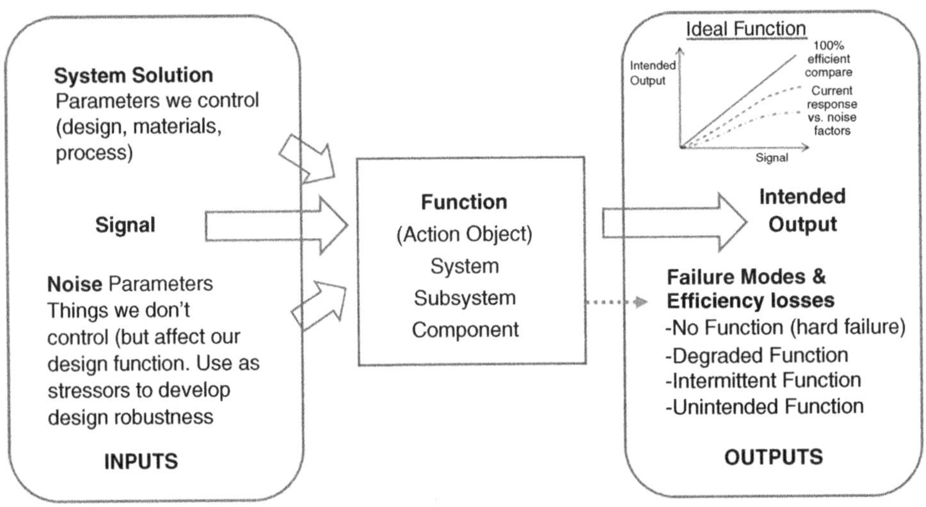

Figure 8.3 P-Diagram Review.

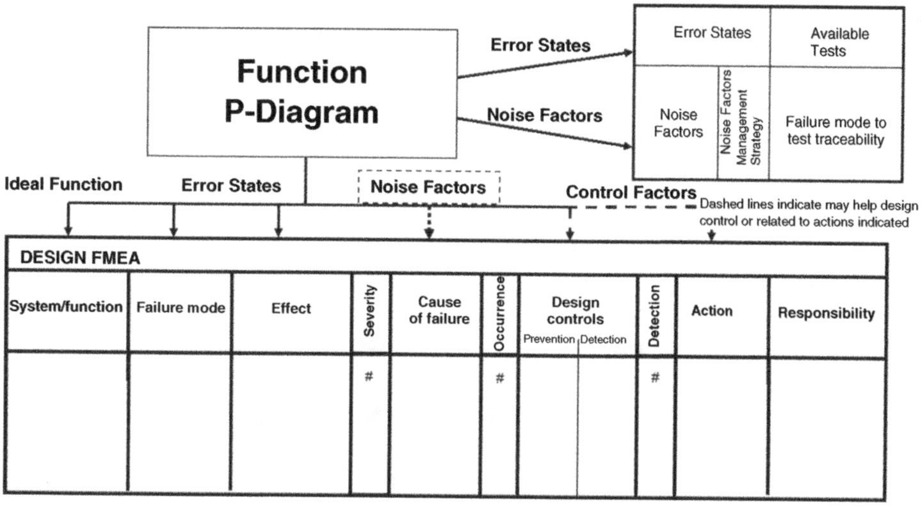

Figure 8.4 P-Diagram, DFMEA, and RRCL Connection.

- Understanding these factors is essential for developing strategies to minimize the impact of noise and enhance the system's robustness.

3) **Optimization of Input–Output Relationships:**
   - By clearly mapping out the input and output flows, the P-diagram aids in optimizing these relationships to ensure consistent and reliable performance.
   - This optimization process is crucial for achieving a robust system that can withstand variations and disturbances.
4) **Development of Robust Design Strategies:**
   - The insights gained from the P-diagram enable the application of robust design techniques, such as Taguchi methods, to fine-tune control factors.
   - These strategies help in desensitizing the system to noise factors, thereby improving its overall reliability and performance.

Serving as an Input for DFMEA shown as in Figure 8.4

1) **Systematic Identification of Failure Modes:**
   - The P-diagram serves as a structured framework for identifying potential failure modes within the system.
   - By analyzing the interactions between inputs, outputs, control factors, and noise factors, potential points of failure can be systematically identified.
2) **Assessment of Failure Effects:**
   - The P-diagram helps in understanding the effects of identified failure modes on the system's performance.
   - This assessment is crucial for prioritizing failure modes based on their severity, occurrence, and detectability, forming a foundation for effective risk management.
3) **Development of Mitigation Strategies:**
   - Using the insights from the P-diagram, robust mitigation strategies can be developed to address potential failure modes.

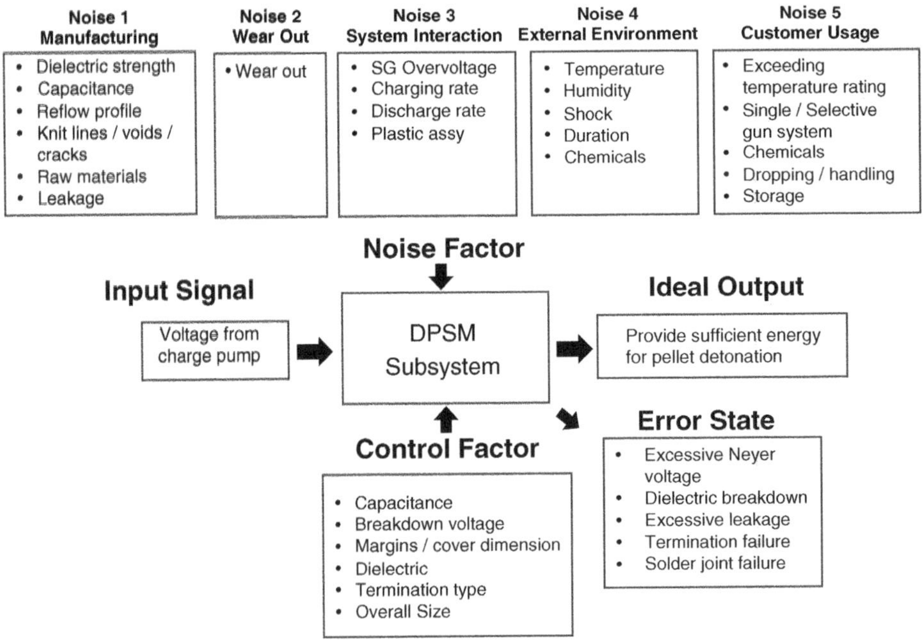

Figure 8.5 DPSM P-Diagram Communication.

- These strategies might include design changes, process improvements, or the implementation of control measures to enhance system reliability.

4) **Enhanced Communication and Documentation as shown in Figure 8.5:**
   - The P-diagram serves as a clear and comprehensive communication tool that can be used by cross-functional teams.
   - It ensures that all stakeholders have a consistent understanding of the system's functional relationships and potential failure points, facilitating collaboration and informed decision-making.
5) **Support for Continuous Improvement:**
   - The P-diagram is a dynamic document that evolves with the system design and development process.
   - It supports ongoing DFMEA efforts by providing a structured approach to continuously identify and address new potential failure modes as they arise.
6) After completing the P-diagram and DFMEA, the next step involves several critical activities below. By following these steps, we can ensure a comprehensive approach to addressing potential failures, improving the robustness of the design against noise factors, and effectively managing noise throughout the design process.

The next step involves several critical activities as shown in Figure 8.6 including:

1) **Mapping and Linking Failure Modes to Noise Factors:**
   - **Identify Noise Factors:** Determine the specific noise factors that contribute to each identified failure mode.
   - **Linking Process:** Create a clear association between the failure modes and their corresponding noise factors.

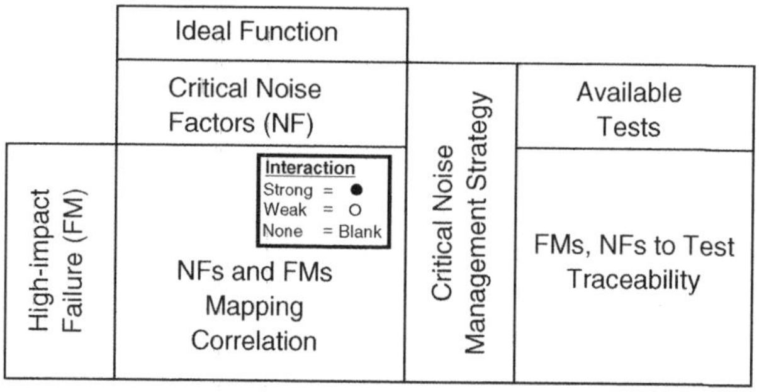

Figure 8.6 FMs, NFs, NFMS, and Tests Mapping.

2) **Assessing Design Verification Tests:**
   - **Test Capability:** Evaluate the existing design verification (DV) tests to determine their effectiveness in addressing the identified noise factors and failure modes.
   - **Gap Analysis:** Identify any gaps where noise factors are not adequately validated by current DV methods.
3) **Revising and Adding Tests:**
   - **Revise Existing Tests:** Modify current DV tests to better address the noise factors and failure modes where gaps have been identified.
   - **Develop New Tests:** Design and implement new tests to cover noise factors and failure modes that are not addressed by existing DV methods.
4) **Developing Noise Management Strategies:**
   - **Concept Design Changes:** Consider altering the overall design concept to mitigate noise factors.
   - **Parameter Design:** Utilize parameter design techniques to optimize design parameters for noise reduction.
   - **New Technologies:** Investigate and implement new technologies that can help manage or eliminate noise factors.
   - **Design Upgrades:** Upgrade existing design elements to enhance their robustness against noise factors.
   - **Other Strategies:** Explore additional methods for managing noise factors, tailored to the specific requirements of the project.
5) **Utilizing the Robustness Checklist (RCL):**
   - **Preparation of RCL:** Develop a robustness checklist (RCL) specifically for this purpose, as shown in Figures 8.7 and 8.8.
   - **Checklist Components:** The RCL is composed of eight components as follows:
     1. **Ideal Function** (*Source:* P-diagram).
     2. **High-Impact Potential Failure Modes** (*Source:* P-diagram/DFMEA).
     3. **Critical Noise Factors:** a) Manufacturing b) Aging c) Customer Usage d) External Environment e) System Interactions.
     4. **High-Impact Potential Failure Modes to Critical Noise Factor Mapping.**
     5. **Noise Factor Management Strategy (NFMS).**
     6. **Available Tests.**
     7. **Error States to Available Tests' Traceability.**
     8. **Noise factors to test traceability.**

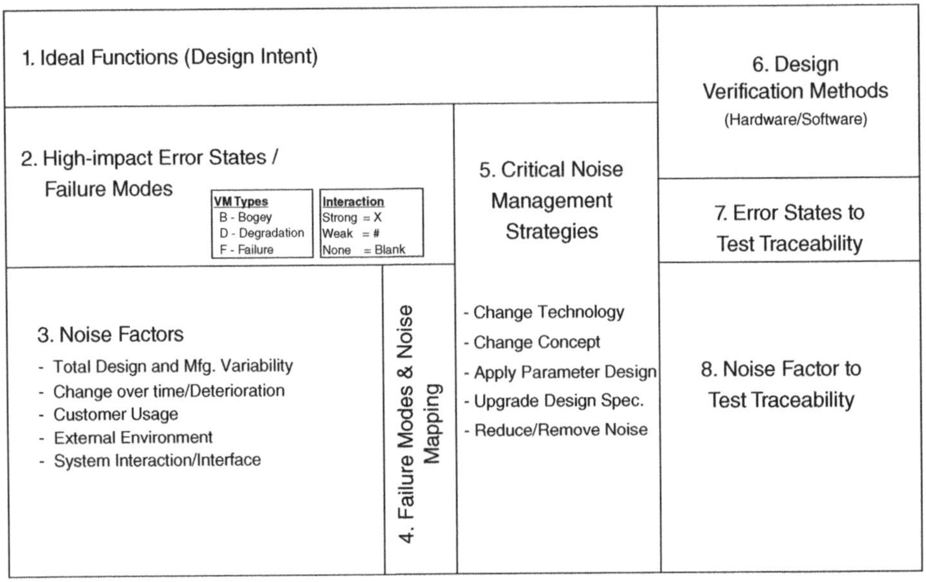

Figure 8.7 DPSM P-RRCL.

Up to this point, a comprehensive approach has been taken to address potential failures, enhance the robustness of the design against noise factors, and manage noise effectively throughout the design process. The next step involves demonstrating the design's capability to perform as intended throughout its useful life, even in the presence of these noise factors. This design intent can be quantified by either meeting a specific performance target or by making comparisons against a baseline. The team should be prepared to visually demonstrate the robustness or improvements once the test results are available, particularly during the demonstration phase.

In addition to achieving optimized performance in a robust manner, it is crucial to develop and implement a **mistake prevention strategy** as part of the overall system optimization process. By proactively addressing potential errors in both operation and manufacturing, the system can be safeguarded against failures that could compromise its reliability and efficiency.

Mistake Prevention Strategy

Mistake prevention focuses on ensuring that both the system and its processes are designed to avoid errors before they occur. This is particularly important in high-stakes environments where even small mistakes can lead to significant failures or disruptions. To achieve this, the **poka-yoke** technique, also known as **mistake-proofing**, plays a key role in preventing errors in both production and usage.

Mistake-Proofing (Poka-Yoke) Techniques

- **Operational Error Elimination:** Poka-yoke techniques aim to eliminate errors that could occur during the operation of the system by making incorrect actions physically impossible or immediately detectable. For example, design features such as foolproof

Program: DPSM Subsystem: DPSM CP8

IDEAL FUNCTIONS

a) Stores sufficient energy for pellet detonation
b) Maintains connection from electrodes to PCB (solder / termination integrity)
c) Fit in package

VM Types	Interaction
B - Bogey	Strong = X
D - Degradation	Weak = #
F - Failure	None = Blank

HIGH IMPACT		ERROR STATES
X	E	Solder joint failure
	D	Termination failure
	C	Leakage
X	B	Dielectric breakdown
X	A	Excessive Neyer voltage

Noise Factors Management Stratigies (NFMS)

I Change technology
II Apply robust parameter design
III Upgrade Design Spec.
IV Reduce / Remove Noise
V Add Compensation Device

VERIFICATION METHODS

ERROR STATES	1 Leakage	2 Thermal cycling	3 Thermal shock	4 Construction analysis / solder joint	5 Neyer test RT + 175C	6 Reflow sensitivity	7 Bending stress	8 BDV at 175C	9 Explosive shock	10 Ramp rate sensitivity	11 BDV (pressure + temp +	
	D	F	F	D	D	D	F	D	B	D	D	VM TYPE
E		x	x	x			x		x			
D		x	x	x			x		x			
C	x	#	#									
B		#	#	#		x		x	x	x	x	
A					x							
	17	46	56	36	25	29	25	10	30	10	37	HIGH IMPACT

Noise	A	B	C	D	E	NFMS Descriptions	1	2	3	4	5	6	7	8	9	10	11
NOISE 1: Manufacturing																	
a) Dielectric strength and …..		x	x			I, III, IV (TRS new dielctric, AVX upg spec, all manu Kpc))		#	#	#		x		x	x	x	x
b) Capacitance and ……	x					I, II, IV (TRS new dielctric, AVX upg spec, all manu Kpc))		x	x		x						
NOISE 2: Wear Out																	
a) Not a significant concern due to mission profile																	
NOISE 3: System Interaction																	
a) SG overvoltage and ……		x				III (upgrade bdv)						x		x			x
NOISE 4: External Environment																	
a) Humidity		x	x			I, III (TRS new dielctric, AVX upg spec)	x	x	x								x
b) Shock and ……		x	x	x	x	III, VI			x				x		x		
NOISE 5: Customer Usage																	
c) Shipping / handling / storage and ….		#		#	#	III	x	x	x				x		x		x

Figure 8.8 Example of DPSM RRCL.

connectors or automated alerts can prevent users from making incorrect choices that might lead to system damage or failure.
- **Manufacturing Error Prevention:** In the manufacturing process, mistake-proofing can ensure that components are assembled correctly by preventing improper alignments or incorrect part placements. Techniques such as **guide pins**, **color coding**, and **sensor integration** can help detect and correct errors before they propagate into larger system issues.

By integrating mistake-proofing methods throughout both design and production, systems can achieve higher levels of reliability and user satisfaction, reducing the likelihood of costly post-production errors or product recalls.

Hard Failure Mitigation

While mistake prevention techniques focus on eliminating operational and manufacturing errors, it is equally important to address **hard failures** that arise from material and mechanical breakdowns. Hard failures occur when a system or component is no longer able to function due to structural damage or material fatigue. To mitigate such failures, design robustness must be ensured at every level.

Ensuring Design Robustness for Hard Failure Mitigation

- **Material Durability:** Select materials that can withstand the environmental stresses the system will encounter during its lifetime. This includes consideration of factors like temperature fluctuations, moisture, chemical exposure, and mechanical stress. Robust material selection reduces the risk of fatigue, cracking, or warping over time.
- **Mechanical Stability:** Ensure that the design is mechanically stable by addressing potential stress points and load-bearing issues. Finite element analysis (FEA) and similar tools can be used to simulate the effects of operational stresses on mechanical components, allowing for the design to be strengthened in critical areas.
- **Redundancy and Fail-safes:** Incorporate redundancy into critical system components to ensure that if one part fails, a backup can take over without causing the entire system to break down. Fail-safe mechanisms such as backup power supplies or automatic shutoffs can prevent hard failures from leading to catastrophic outcomes.

By focusing on both mistake-proofing techniques and hard failure mitigation, a comprehensive mistake prevention strategy can be established that not only prevents errors before they occur but also ensures that the system can withstand material and mechanical failures.

Key Elements of the Mistake Prevention Strategy

- **Operational and Manufacturing Error Elimination:** Through mistake-proofing techniques, the system is designed to prevent errors at both the operational and manufacturing stages. This ensures greater reliability and minimizes the potential for human error to negatively impact system performance.
- **Hard Failure Mitigation:** By ensuring design robustness, the system is protected against material and mechanical breakdowns, allowing for sustained performance and extended product life.

8.5 Conclusion: Comprehensive Approach to Robust Optimization and Mistake Prevention

A robust optimization process must account not only for performance variability due to noise factors but also for potential operational and manufacturing errors that could lead to system failure. By incorporating a well-structured mistake prevention strategy, including mistake-proofing techniques and hard failure mitigation, the system can be designed to be highly reliable and resilient.

This combined approach ensures that the system performs optimally under a range of conditions, with mechanisms in place to prevent both minor operational errors and major mechanical failures. As a result, the final product is not only robust in its performance but also durable, reliable, and protected against a wide variety of risks.

8.5.1 Verify Phase Purpose in Robust Design for Reliability

The Verify phase is crucial for ensuring that the product design meets all reliability and performance requirements before full-scale production begins. This phase ensures the final design aligns with the product development process, meets customer expectations, and adheres to quality and reliability standards. This thorough verification process is essential for delivering a reliable, high-quality product to the market.

The Verify phase validates that the design is robust and performs consistently under various conditions, addressing all identified risks and variabilities [5]. In addition to the common approach as mentioned in Chapter 4, our featured approach strengthens the Verify phase by focusing on the major deliverables in the scorecard shown as in Table 8.6.

The robustness demonstration matrix (RDM) as shown in Figures 8.9 and 8.10 is constructed in a systematic way to provide a structured approach for evaluating the effectiveness of noise factor management strategies [1]. By organizing and documenting the interactions between design variables and potential noise factors, the RDM enables teams to assess whether their strategies are mitigating these factors as intended. Without this matrix, it becomes challenging to determine if the efforts to control noise factors are genuinely effective or if they merely provide temporary solutions. Additionally, the RDM plays a crucial role in demonstrating improvements over an established baseline, offering clear, data-driven evidence that the design is more robust and resilient against variability. This systematic approach ensures that improvements are not only made but also clearly documented and validated, providing confidence in the design's long-term performance.

The elements of the RDM are shown in Figure 8.10.

VM requirements are performance requirements that

- Help evaluate the occurrence of error states/achievement of objectives.
- Use engineering metrics to characterize a failure mode or function during a test.
- Act as a reference point against which improvements can be assessed.
- Can be viewed as a threshold for failure mode, and a distance from the threshold is a measure of robustness.

Table 8.6 Verify (V) Phase Scorecard.

Verify (V)	**Gate 5: Validation and Deployment - Avoid failure modes due to unrepresen-tative tests**	Ensure the product is properly tested for robustness and mistake-proofing Validation ensures the product can perform under all expected conditions	• **Design Verification Report:** Document final testing results, including results from upgraded tests that reflect real-world use cases- **Reliability Demonstration Matrix (RDM):** Use the RDM to ensure the product has been tested against all potential failure modes • **Robustness Assessment:** Evaluate the product's ability to maintain performance under varying conditions and identify areas for robustness improvement. • **Design Review based on Failure Modes (DRBFM):** Conduct thorough reviews to identify both soft and hard failure modes • **Final Design Review based on Test Results (DRBTR):** Perform a comprehensive final review to address any remaining failure modes • **Update Robustness and Reliability Demonstration Matrix:** Reflect the latest test results and design changes to ensure all reliability and robustness targets are met • **Verify and Update CTQs/CTRs for Critical Components and/or Subsystems:** Review and update critical-to-quality (CTQ) and critical-to-reliability (CTR) metrics for key components and subsystems, incorporating risk assessments to identify high-risk functions • **Develop or Update Critical Parameter Management (CPM) Plan:** Establish or revise the CPM plan to manage key parameters critical to the product's reliability and performance • **Identify and document Health Indicators for PHM based on CTQ and CTR**

We want to highlight the importance of expressing VM targets in engineering units rather than in terms of cycles or elapsed time. Examples of engineering units include torque, flow, pressure, and speed fluctuation. A VM target should clearly convey both the metric and its specific value. For instance, speed fluctuation is the metric, and ±5 rpm is the metric value. If the error state is already expressed in engineering terms, only the metric value is necessary. When failure modes are described using engineering metrics, they provide valuable

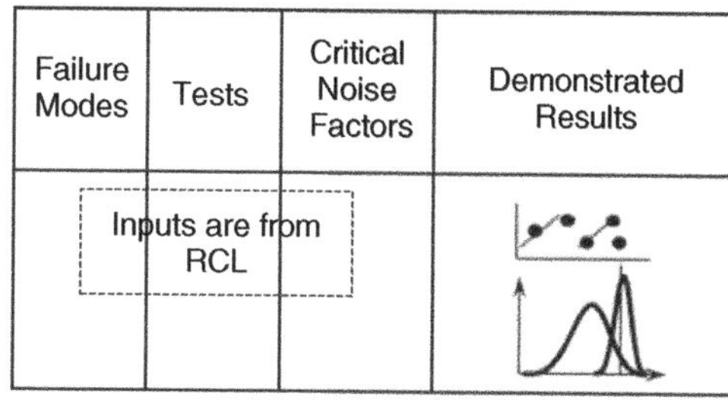

Figure 8.9 Robustness Demonstration Matrix.

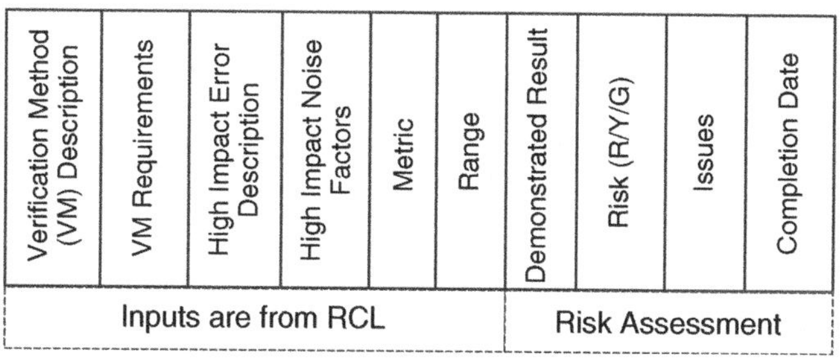

Figure 8.10 Elements of RDM.

insights into the underlying failure mechanisms. A detailed RDM example is shown in Figure 8.11.

What do you do with RDM Results?

- Demonstrate the effectiveness of design actions (implemented noise factor management strategies).
- Quantify robustness improvement graphically.
- Assess risk (R/Y/G) for each RDM candidate based on test results.
- Document issues associated with demonstrated results and develop an Issue resolution plan, if risk exists.

Key Areas to Consider:

- Alignment with High-Impact Error States: Are the candidates for the robustness demonstration matrix (RDM) directly linked to the high-impact error states identified in the robustness checklist (RCL)?
- Effectiveness of Testing Strategy: Is the testing strategy adequately designed to demonstrate meaningful improvements in the design?
- Monitoring Functional Degradation: Have you considered measuring functional degradation at various intervals throughout the testing process, rather than only at the beginning and end?
- Capturing Interactions Among Noise Factors: Does the sequencing of existing tests effectively capture all significant interactions among noise factors?
- Appropriate Level of System Hierarchy for Testing: Are the tests being conducted at the appropriate level of the system hierarchy to ensure accurate results?

System	Assy Subassy/ Suppliers	High Impact Failure Mode(s)	Verification Method / Test	High Impact Noise Factors	Reliability Requirement (Cycles, Time, etc.)	Test Target	Acceler -ation Model	Demonstrated Result	Risk (R/Y/G)	Issues	Completion Date
DPSM	CP8	A	NT Test See new VMs	See RCL for details	See test requirements for details	See test requirements for details	None	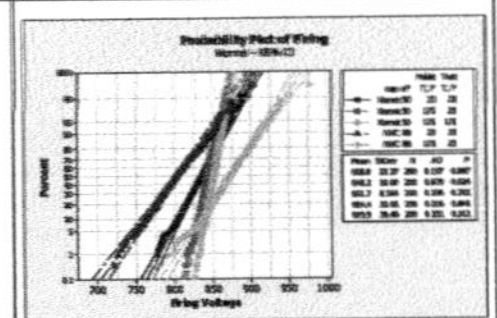	G	None	
DPSM	CP8	B, C, D, E	Thermal shock Refer to new VMs	See RCL for details	See test requirements for details	See test requirements for details	AMTS	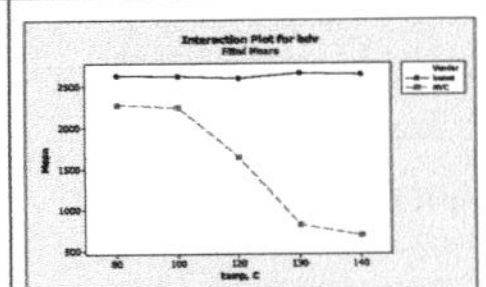	G	None	
DPSM	CP8	B and C	Humidity	See reference 1	See test requirements for details	See test requirements for details	AMH	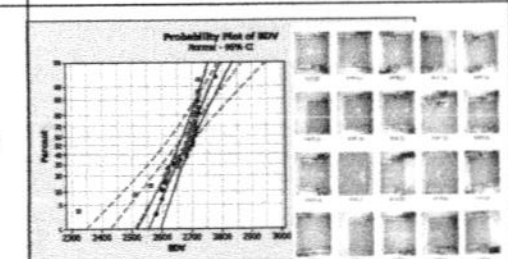	G	None	

Figure 8.11 Example of RRDM for DPSM.

- Reflecting Noise Factor Management Strategies in Test Parts: Do the test parts accurately reflect the actions taken in the noise factor management strategy?
- Engineering Units for Verification Metrics (VM): Can the verification metrics (VM) targets be expressed in engineering units for clarity and precision?
- Baseline for Comparison: Is there a baseline available to clearly demonstrate improvements made through the design process?
- Engineering Confidence: Upon completion, will the RDM provide you with the necessary engineering confidence in the design's robustness and reliability?

Summary of Benefits for the Robust Design for Reliability (RDfR) Case Study on the Detonator Power Supply Module (DPSM)

The DPSM case study provides an excellent real-world example of how the RDfR framework can be successfully implemented to improve product performance, reliability, and manufacturing processes.

Demonstrated RDfR-Driven Advancement in Lifecycle Reliability, Six Sigma Performance, and System-Level Robustness

Figures 8.12 and 8.13 illustrate the enhanced performance of the redesigned Detonator Power Supply Module (DPSM) following implementation of the Robust Design for Reliability (RDfR) methodology. The updated design demonstrates significantly improved output stability, reduced performance variation, and increased resistance to environmental and manufacturing noise factors.

In addition, Figure 8.13 presents the corresponding manufacturing process capability analysis. The post-RDfR results show a process capability index of **Cpk = 2.18**, indicating a highly capable and statistically controlled process with substantial margin relative to specification limits. The narrow within-process variation and strong alignment between within and overall distributions confirm that variability has been effectively minimized at both the design and production levels.

A Cpk value of 2.18 corresponds to world-class manufacturing performance, exceeding traditional Six Sigma benchmarks (Cpk ≈ 2.0). This level of capability reflects not only improved mean centering but also a significant reduction in process dispersion relative to

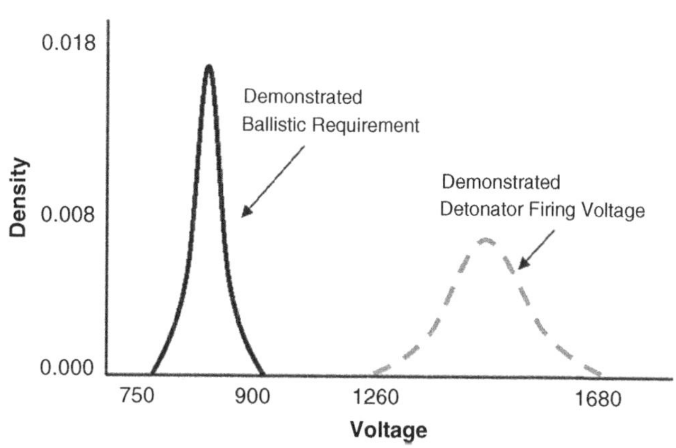

Figure 8.12 DPSM New Design Performance.

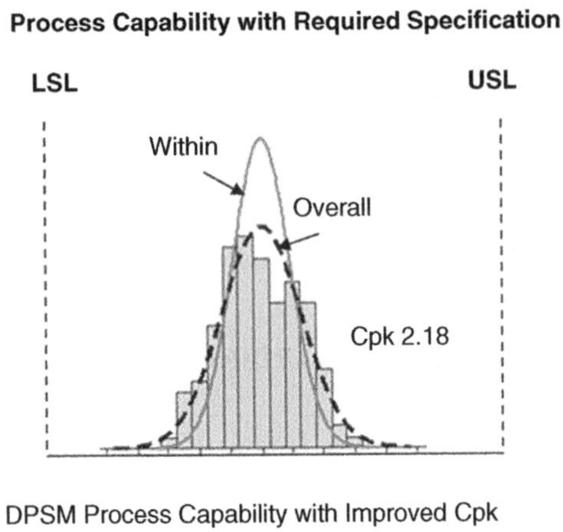

Figure 8.13 Improved Process Capability.

specification boundaries, resulting in an extremely low probability of out-of-specification product during production and field operation.

Collectively, these results validate that the RDfR approach enhanced product functional robustness while simultaneously elevating manufacturing consistency, thereby ensuring reliable field performance and long-term lifecycle stability.

Let us expand on the key aspects and outcomes of this case study:

Implementation of RDfR Framework

The DPSM case study demonstrates the practical application of the RDfR framework, which integrates robustness thinking with Design for Six Sigma (DFSS) principles. This approach addresses both the conceptual ("What") and practical ("How To") aspects of product development, with a strong emphasis on reliability.

Data-driven Decision-making

The team leveraged robust optimization studies to enhance their decision-making processes. This involved:

- Collecting and analyzing extensive data on DPSM performance.
- Identifying critical parameters affecting reliability and safety.
- Using statistical tools to model and predict outcomes.
- Conducting sensitivity analyses to understand the impact of various factors.

By basing decisions on solid data and statistical analysis, the team was able to make more informed choices throughout the development process.

Optimization in Uncontrollable User Environments

A crucial aspect of the DPSM case study was the optimization of the module in the presence of uncontrollable user environments. This step was essential in building engineering confidence and ensuring the module's robust performance under real-world conditions.

Importance of Optimization

- **Real-World Variability:** Products like the DPSM are often used in diverse and unpredictable environments. Factors such as temperature fluctuations, humidity, and varying power supply conditions can significantly impact performance.

- **Robust Design:** By optimizing the DPSM to perform reliably despite these uncontrollable factors, the design team ensured that the product would meet its performance and reliability targets in any user environment.

Building Engineering Confidence

- **Predictive Modeling:** The team used advanced modeling techniques to predict how the DPSM would behave under various conditions. This involved simulating different environmental scenarios and stress factors.
- **Robustness Testing:** By subjecting the DPSM to rigorous robustness testing, the team was able to identify potential failure modes and address them proactively.
- **Confidence in Performance:** This thorough optimization process gave engineers the confidence that the DPSM would perform reliably in the field, reducing the risk of unexpected failures.

Rigorous Testing and Validation

With the confidence gained from the optimization phase, the team proceeded to conduct rigorous testing and validation of the DPSM under various conditions:

- **Comprehensive Testing Protocols:** The DPSM was subjected to a wide range of tests, including thermal cycling, vibration testing, and electrical stress tests, to ensure it could withstand real-world conditions.
- **Validation Across Scenarios:** The module was tested in different scenarios to validate its performance across the entire spectrum of potential user environments.
- **Feedback Loop:** Data collected during testing was used to further refine the design, creating a feedback loop that continuously improved the product's reliability and performance.

By optimizing the DPSM for uncontrollable user environments and conducting rigorous testing and validation, the team ensured that the module would deliver consistent, reliable performance in any situation. This approach not only enhanced the product's robustness but also provided a solid foundation for its successful deployment in the field.

Reliability and Safety Improvements

One of the most significant outcomes was the achievement of 100% reliability and safety performance. This was accomplished through:

- Rigorous testing and validation of the DPSM under various conditions based on the well-constructed Reliability and Robustness Demonstration Matrix (RRDM).
- Continuous monitoring and feedback systems.
- Strict quality control measures throughout the manufacturing process.

Performance Optimization

The case study highlights how the team met performance targets while achieving zero defects:

- Precise calibration of DPSM components.
- Optimization of power output and stability.
- Implementation of advanced control algorithms.
- Extensive testing to ensure consistent performance across all units.

Manufacturing Process Enhancements

A key achievement was the significant improvement in manufacturing process capability:

- Cpk values increased from 0.24 to 2.18, indicating a much tighter control over the manufacturing process.
- This improvement led to more consistent product quality and fewer variations between units.
- Implementation of statistical process control (SPC) techniques.
- Continuous monitoring and adjustment of manufacturing parameters.

Cost Reduction

The case study demonstrates substantial cost savings:

- Annual manufacturing costs were reduced by $340,000.
- This was achieved through:
  - Streamlined production processes.
  - Reduced waste and rework.
  - Improved supply chain management.
  - Optimization of material usage.

Practical Application of RDfR

The DPSM case study serves as a roadmap for implementing RDfR strategies in other projects:

1) **Problem Identification:** Clearly define the reliability and performance issues to be addressed.
2) **Data Collection:** Gather comprehensive data on current performance and potential factors affecting reliability.
3) **Analysis:** Use statistical tools to analyze data and identify key parameters.
4) **Optimization:** Apply robust optimization techniques to improve design and manufacturing processes.
5) **Implementation:** Introduce changes in a controlled manner, monitoring results closely.
6) **Validation:** Rigorously test and validate improvements to ensure they meet or exceed targets.
7) **Continuous Improvement:** Establish processes for ongoing monitoring and refinement.

By following this approach, other projects can potentially achieve similar improvements in reliability, performance, and cost effectiveness. The DPSM case study clearly illustrates how the integration of robustness thinking and DFSS principles can lead to significant advancements in product development and manufacturing processes, ultimately resulting in more reliable, safer, and cost-effective products.

References

1 Hu, M. (2014). Robustness Thinking in Design for Reliability, RAMS, 2014 Annual RELIABILITY and MAINTAINABILITY Symposium.

2 Taguchi, G., Chowdhury, S., and Yuin, W. (2004). *Taguchi Quality Engineering Handbook*, 1449. Hoboken, New Jersey, USA: John Wiley & Sons, Inc. (Chapter 41, Page 1449, authored by Matthew Hu & others.

3 Clausing, D. (1994). *Total Quality Development. A Step-By-Step Guide to World-Class Concurrent Engineering*. New York: ASME Press.

4 Readiness Plan, Transfer function-based design to improve product reliability and robustness in design for six sigma, Matthew Hu, Six Sigma Forum Magazine, American Society for Quality, August 2013.

5 Matthew, H., Pieprzak, J.M., and Glowa, J. (2004). *Essentials of Design Robustness in Design for Six Sigma (DFSS) Methodology*, 13. Warrendale, Pennsylvania, USA: SAE International, SAE Technical Paper Number: 2004-01-0813.

6 Matthew, H. and Yang, K. (2005). *Transfer Function Development in Design for Six Sigma Framework*. Warrendale, Pennsylvania, USA: SAE International.

7 Hu, M., Yang, K., and Taguchi, S. (2000, October). *Enhancing robust design with the aid of TRIZ and axiomatic design (Part I)*. In Proceedings of the Axiomatic Design Conference; also published in The TRIZ Journal.

8 Matthew, H. and Pieprzak, J.M. (2005). Using axiomatic design to improve conceptual design robustness in design for six sigma (DFSS) methodology. *International Journal of Six Sigma and Competitive Advantage* 1 (3): 245–262.

9 Taguchi, G., Chowdhury, S., and Taguchi, S. (2000). *Robust Engineering: Learn How to Boost Quality While Reducing Costs & Time to Market*. New York, New York, USA: Mc Graw Hill.

Index

n

o

p

q

r

s

Printed and bound by CPI Group (UK) Ltd, Croydon, CR0 4YY
17/09/2026
14964270-0002